VOLUME ONE HUNDRED AND FIVE

ADVANCES IN PROTEIN CHEMISTRY AND STRUCTURAL BIOLOGY

Insights into Enzyme Mechanisms and Functions from Experimental and Computational Methods

VOLUME ONE HUNDRED AND FIVE

ADVANCES IN PROTEIN CHEMISTRY AND STRUCTURAL BIOLOGY

Insights into Enzyme Mechanisms and Functions from Experimental and Computational Methods

Edited by

CHRISTO Z. CHRISTOV

Faculty of Health and Life Sciences,
Northumbria University,
Newcastle upon Tyne,
United Kingdom

ELSEVIER

AMSTERDAM • BOSTON • HEIDELBERG • LONDON
NEW YORK • OXFORD • PARIS • SAN DIEGO
SAN FRANCISCO • SINGAPORE • SYDNEY • TOKYO
Academic Press is an imprint of Elsevier

Academic Press is an imprint of Elsevier
50 Hampshire Street, 5th Floor, Cambridge, MA 02139, United States
525 B Street, Suite 1800, San Diego, CA 92101-4495, United States
The Boulevard, Langford Lane, Kidlington, Oxford OX5 1GB, United Kingdom
125 London Wall, London, EC2Y 5AS, United Kingdom

First edition 2016

Notices
Knowledge and best practice in this field are constantly changing. As new research and experience broaden our understanding, changes in research methods, professional practices, or medical treatment may become necessary.

Practitioners and researchers must always rely on their own experience and knowledge in evaluating and using any information, methods, compounds, or experiments described herein. In using such information or methods they should be mindful of their own safety and the safety of others, including parties for whom they have a professional responsibility.

To the fullest extent of the law, neither the Publisher nor the authors, contributors, or editors, assume any liability for any injury and/or damage to persons or property as a matter of products liability, negligence or otherwise, or from any use or operation of any methods, products, instructions, or ideas contained in the material herein.

ISBN: 978-0-12-804825-2
ISSN: 1876-1623

Publisher: Zoe Kruze
Acquisition Editor: Alex White
Editorial Project Manager: Helene Kabes
Production Project Manager: Surya Narayanan Jayachandran
Cover Designer: Mark Rogers

Typeset by SPi Global, India

CONTENTS

Contributors *vii*
Preface *ix*

1. Molecular Modeling and Chemoinformatics to Advance the Development of Modulators of Epigenetic Targets: A Focus on DNA Methyltransferases **1**

F.D. Prieto-Martínez, A. Peña-Castillo, O. Méndez-Lucio, E. Fernández-de Gortari, and J.L. Medina-Franco

1. Introduction 2
2. Progress on Chemical Information 4
3. Chemoinformatic Studies of DNMTs 9
4. VS: Hit Identification and Optimization 16
5. Computer-Assisted Drug Repurposing 19
6. Food Chemicals as Potential Modulators of DNMTs and Other Epigenetic Targets 20
7. Concluding Remarks 21

Acknowledgments 22
References 23

2. Recent Advances in Computational Models for the Study of Protein–Peptide Interactions **27**

D. Kilburg and E. Gallicchio

1. Introduction 28
2. Theory and Methods 30
3. Sample Applications 39
4. Concluding Remarks 51

References 51

3. Structural Plasticity in Globins: Role of Protein Dynamics in Defining Ligand Migration Pathways **59**

C. Estarellas, L. Capece, C. Seira, A. Bidon-Chanal, D.A. Estrin, and F.J. Luque

1. Introduction 60
2. Ligand Migration Pathways in the Globin Superfamily: Structure–Function Relationships 63
3. The Coordination of the Heme Group and Its Role in Ligand Binding 73

4. Conclusions 76
Acknowledgments 77
References 77

4. Dynamic Reorganization and Enzymatic Remodeling of Type IV Collagen at Cell–Biomaterial Interface 81

N.M. Coelho, V. Llopis-Hernández, M. Salmerón-Sánchez, and G. Altankov

1. Introduction 82
2. Material and Methods 85
3. Results and Discussion 91
4. Conclusive Remarks 100

References 101

5. Cytochromes P450: History, Classes, Catalytic Mechanism, and Industrial Application 105

D.J. Cook, J.D. Finnigan, K. Cook, G.W. Black, and S.J. Charnock

1. Background and History of Cytochromes P450 106
2. P450 Classes 110
3. P450 Family Nomenclature 114
4. Catalytic Cycle 115
5. Uncoupling 117
6. P450s in Industry 119

References 123

6. Understanding the Multifaceted Role of Human Down Syndrome Kinase DYRK1A 127

L.J. Kay, T.K. Smulders-Srinivasan, and M. Soundararajan

1. Introduction 128
2. Minibrain in *Drosophila* 129
3. Role of DYRK1A in Infection and Immunity 132
4. Cardiovascular Function of DYRK1A 136
5. Role of DYRK1A in Cancer 139
6. Understanding the Neuronal Function of DYRK1A 148
7. Concluding Remarks 157

References 158

Author Index 173
Subject Index 195

CONTRIBUTORS

G. Altankov
Institute for Bioengineering of Catalonia (IBEC); Institució Catalana de Recerca i Estudis Avançats (ICREA), Barcelona, Spain

A. Bidon-Chanal
Institute of Biomedicine (IBUB), Faculty of Pharmacy and Food Science, University of Barcelona, Santa Coloma de Gramenet, Spain

G.W. Black
Faculty of Health and Life Sciences, Northumbria University, Newcastle upon Tyne, United Kingdom

L. Capece
Universidad de Buenos Aires, Facultad de Ciencias Exactas y Naturales, INQUIMAE-CONICET, Ciudad Universitaria, Ciudad de Buenos Aires, Argentina

S.J. Charnock
Prozomix Limited, Haltwhistle, Northumberland, United Kingdom

N.M. Coelho
Institute for Bioengineering of Catalonia (IBEC), Barcelona, Spain

D.J. Cook
Prozomix Limited, Haltwhistle, Northumberland; Faculty of Health and Life Sciences, Northumbria University, Newcastle upon Tyne, United Kingdom

K. Cook
Thermo Fisher Scientific, Hemel Hempstead, Hertfordshire, United Kingdom

C. Estarellas
Institute of Biomedicine (IBUB), Faculty of Pharmacy and Food Science, University of Barcelona, Santa Coloma de Gramenet, Spain

D.A. Estrin
Universidad de Buenos Aires, Facultad de Ciencias Exactas y Naturales, INQUIMAE-CONICET, Ciudad Universitaria, Ciudad de Buenos Aires, Argentina

E. Fernández-de Gortari
Facultad de Química, Universidad Nacional Autónoma de México, México City, México

J.D. Finnigan
Prozomix Limited, Haltwhistle, Northumberland; Faculty of Health and Life Sciences, Northumbria University, Newcastle upon Tyne, United Kingdom

E. Gallicchio
Brooklyn College, Brooklyn; The Graduate Center of the City University of New York, New York, NY, United States

L.J. Kay
Faculty of Health and Life Sciences, Northumbria University, Newcastle upon Tyne, United Kingdom

D. Kilburg
Brooklyn College, Brooklyn; The Graduate Center of the City University of New York, New York, NY, United States

V. Llopis-Hernández
Center for Biomaterials and Tissue Engineering, Universidad Politécnica de Valencia, Valencia, Spain; School of Engineering, University of Glasgow, Glasgow, United Kingdom

F.J. Luque
Institute of Biomedicine (IBUB), Faculty of Pharmacy and Food Science, University of Barcelona, Santa Coloma de Gramenet, Spain

J.L. Medina-Franco
Facultad de Química, Universidad Nacional Autónoma de México, México City, México

O. Méndez-Lucio
Facultad de Química, Universidad Nacional Autónoma de México, México City, México

A. Peña-Castillo
Facultad de Química, Universidad Nacional Autónoma de México, México City, México

F.D. Prieto-Martínez
Facultad de Química, Universidad Nacional Autónoma de México, México City, México

M. Salmerón-Sánchez
School of Engineering, University of Glasgow, Glasgow, United Kingdom

C. Seira
Institute of Biomedicine (IBUB), Faculty of Pharmacy and Food Science, University of Barcelona, Santa Coloma de Gramenet, Spain

T.K. Smulders-Srinivasan
Faculty of Health and Life Sciences, Northumbria University, Newcastle upon Tyne, United Kingdom

M. Soundararajan
Faculty of Health and Life Sciences, Northumbria University, Newcastle upon Tyne, United Kingdom

PREFACE

Enzymes play a crucial role in molecular processes in living organisms. They are characterized with the ability to increase considerably the rate of chemical reactions in the cells but also with high specificity toward their substrates. The key functions of enzymes in cell biology and unique properties make them important targets for drug design and applications in biotechnology, protein engineering, and synthetic biology.

Recent developments in experimental methods contribute for a rapid growth in our understanding about structure–function relationships in enzymes. Structural methods such as X-ray crystallography and NMR are currently widely used to determine the three-dimensional structures of enzymes, and their complexes with substrates and inhibitors. Homology modeling methods received growing applications to predict 3D structures of enzymes, when experimental structures are not available. In addition, spectroscopic and kinetic methods are broadly applied to complement structural information and understand enzyme structure–function relationships.

The applications of computational methods to enzyme mechanisms of functions received a large increase over the last years. The growth in computational power, the development of parallel supercomputers, and GPUs made possible modeling studies to be performed for very large biomolecular systems, containing hundreds of thousands of atoms. Nowadays, we can routinely perform molecular dynamics in order to understand the conformational flexibility of enzyme–ligand complexes in microsecond timescales, to apply combined quantum mechanical and molecular mechanical methods to explore the reaction mechanisms and the origins of the catalytic power in enzymes and to implement methods for prediction of the binding orientations and free energies of enzyme–ligand complexes.

The present thematic volume of *Advances in Protein Chemistry and Structural Biology* is aimed to address some of the recent developments in enzymes. The selected contributions are focused on state-of-the-art applications of computational and experimental methods to understand different aspects of structure–function relationships and mechanisms of enzymes and their applications in biotechnology, cell biology, drug design, and biomaterials.

Christo Z. Christov
Faculty of Health and Life Sciences, Northumbria University,
Newcastle upon Tyne, United Kingdom

CHAPTER ONE

Molecular Modeling and Chemoinformatics to Advance the Development of Modulators of Epigenetic Targets: A Focus on DNA Methyltransferases

F.D. Prieto-Martínez, A. Peña-Castillo, O. Méndez-Lucio, E. Fernández-de Gortari, J.L. Medina-Franco[1]
Facultad de Química, Universidad Nacional Autónoma de México, México City, México
[1]Corresponding author: e-mail addresses: medinajl@unam.mx; jose.medina.franco@gmail.com

Contents

1. Introduction 2
2. Progress on Chemical Information 4
3. Chemoinformatic Studies of DNMTs 9
   3.1 Characterization of Chemical Space: ERCS 9
   3.2 Chemoinformatic-Based Pharmacophore Model 12
   3.3 Activity Landscape Modeling 14
   3.4 Quantitative Structure–Activity Relationships 15
4. VS: Hit Identification and Optimization 16
   4.1 Novel VS Hits 16
   4.2 Follow-Up of VS Hits 18
5. Computer-Assisted Drug Repurposing 19
6. Food Chemicals as Potential Modulators of DNMTs and Other Epigenetic Targets 20
7. Concluding Remarks 21
Acknowledgments 22
References 23

Abstract

In light of the emerging field of *Epi-informatics*, ie, computational methods applied to epigenetic research, molecular docking, and dynamics, pharmacophore and activity landscape modeling and QSAR play a key role in the development of modulators of DNA methyltransferases (DNMTs), one of the major epigenetic target families. The increased chemical information available for modulators of DNMTs has opened up the avenue to explore the epigenetic relevant chemical space (ERCS). Herein, we discuss

Advances in Protein Chemistry and Structural Biology, Volume 105
ISSN 1876-1623
http://dx.doi.org/10.1016/bs.apcsb.2016.05.001

recent progress on the identification and development of inhibitors of DNMTs as potential epi-drugs and epi-probes that have been driven by molecular modeling and chemoinformatics methods. We also survey advances on the elucidation of their structure–activity relationships and exploration of ERCS. Finally, it is illustrated how computational approaches can be applied to identify modulators of DNMTs in food chemicals.

1. INTRODUCTION

DNA methylation is one of the most important epigenetic regulation mechanisms and it is mediated primarily by the family of enzymes DNA methyltransferases (DNMTs). This process involves the addition of a methyl group at the C-5 position of a DNA cytosine residue by the cofactor *S*-adenosyl methionine (SAM). This family of enzymes is formed by DNMT1, DNMT3A, DNMT3B, and DNMT3L. DNMT1 is associated with the maintenance of methylation patterns in DNA. DNMT3A and DNMT3B are de novo DNMTs and are able to transfer a methyl group in nonmethylated CpG's islands. DNMT3L is related to DNMT3A and DNMT3B enhancing their activity (Robertson, 2001).

DNMTs are promising epigenetic targets for the treatment of a number of diseases. For instance, DNA hypermethylation is related to cancer metastasis by silencing the expression genes linked to cell division. DNMTs are implicated in autoimmune diseases and inherited disorders (Gros et al., 2012) and are also promising molecular targets for the treatment of other chronic and degenerative diseases such as Alzheimer's and psychiatric conditions (Gros et al., 2012), and diabetes (Arguelles, Meruvu, Bowman, & Choudhury, 2016). As such, inhibitors of DNMTs (DNMTis) are attractive compounds to be developed as clinical candidates for diverse diseases either alone or as part of combination therapies. Moreover, the development of DNMTis as *epi-probes* is also attractive for the further biochemical understanding of the role of DNMTs in epigenetics. In fact, there are not known good molecular probes for DNMTs although selective probe molecules have been developed for other epigenetic targets (Arrowsmith et al., 2015).

Thus far, 5-azacytidine and decitabine (Fig. 1) are DNMTis that have been approved by the USA Food and Drug Administration—FDA—for the treatment of myelodysplastic syndrome (Derissen, Beijnen, & Schellens, 2013). However, these two compounds are prodrugs that need to be incorporated into DNA. Their high toxicity and low specificity have prompted the search for specific inhibitors preferably of the nonnucleosidic type. To date, several nonnucleosidic inhibitors have been reported which

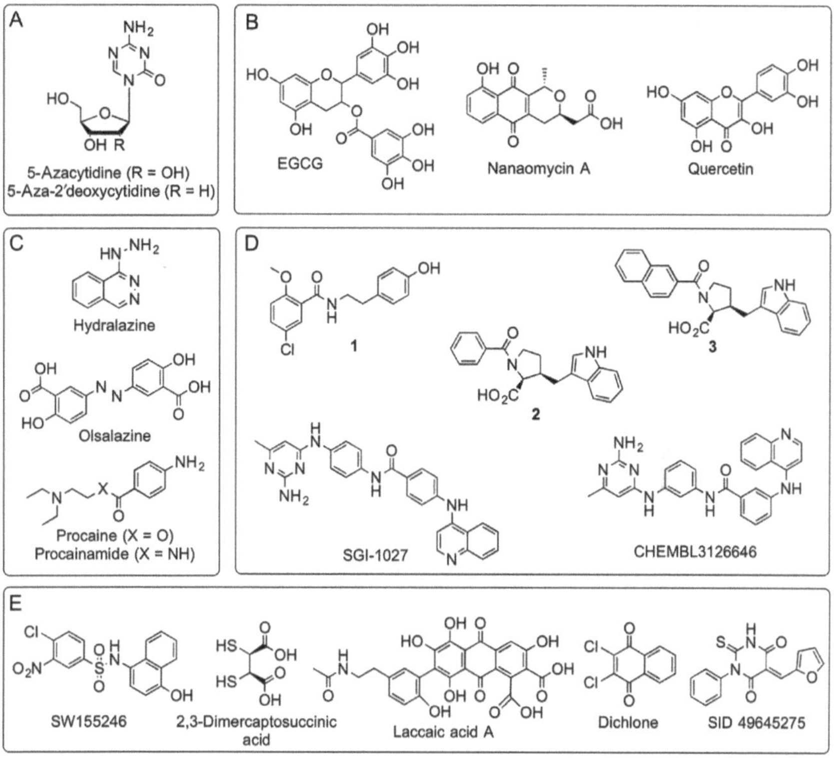

Fig. 1 Representative DNMT inhibitors and compounds with proposed demethylating properties. Compounds are classified by their source: (A) approved for clinical use, (B) natural products, (C) drugs approved for other indications, (D) synthetic compounds coming from optimization programs, and (E) molecules obtained from high-throughput screening.

have been identified from diverse sources such as drugs approved for other indications, natural products, virtual (in silico) screening, high-throughput screening, and synthetic compounds including molecules initially identified from computational screening and later optimized using medicinal chemistry approaches. The chemical structures of representative compounds are shown in Fig. 1. In the figure, compounds are classified by their source in five major groups, namely; DNMTis approved for clinical use (group A), natural products (B), drugs approved for other indications (C), synthetic compounds coming from optimization programs (D), and molecules obtained from high-throughput screening (E). The reader is also referred to reviews of DNMTis that have been published (Erdmann, Arimondo, & Guianvarc'h, 2016; Erdmann, Halby, Fahy, & Arimondo, 2014; Guianvarc'h & Arimondo, 2014).

Computational approaches have played a major role in the identification, optimization, and understanding of the biological activity of DNMTis at the molecular level. Application of in silico techniques continues to increase not only for DNMTs but also for several other epigenetic targets as reflected in the emerging research field called *Epi-informatics* (Medina-Franco, 2016). Reviews on the progress of computational approaches applied to DNMT have been published (Medina-Franco, Méndez-Lucio, Yoo, & Dueñas, 2015; Yoo & Medina-Franco, 2012). However, since these last reviews, major contributions from computational applications have been published in subsequent studies. In particular, a significant and relevant amount of structure–activity information (SAI) has been released and stored in public databases. Herein, we review recent advances on the chemical information resources that have been made available recently to chart the chemical space of DNMTs and elucidate the SAR of DNMTis. We also review major progress on the identification and optimization of hit compounds as well as on the computer-aided discovery of novel hits with putative novel binding sites. All together, these studies emphasize the continued synergy between molecular modeling and chemoinformatic to further advance epigenetic drug and probe development. Representative computational studies, techniques, and major outcomes discussed throughout this chapter are summarized in Table 1.

2. PROGRESS ON CHEMICAL INFORMATION

The SAI of compounds tested as DNMTis has been growing in the last few years. In a recent study, Fernández-de Gortari et al. surveyed public resources and collected chemical databases that contain SAI of DNMTis. Major compound collections included human epigenetic enzyme and modulator database (HEMD) (Huang et al., 2012), ChEMBL (Gaulton et al., 2012), and Binding Database (Liu, Lin, Wen, Jorissen, & Gilson, 2007). An additional database is EpiDBase (Loharch et al., 2015), a curated epigenetic database that includes 11,422 small molecules with activity against different epigenetic targets.

In addition to the availability of molecular databases with experimental data coming from research publications, advances in assay developments and rapid screening technologies for DNMTs (Medina-Franco, Yoo, & Dueñas-Gonzalez, 2015) are facilitating the access to medium- and high-throughput screening data. Table 2 summarizes the information of representative

Table 1 Representative Computational Studies on DNMT Is Recently Reported

Study	Approach	Major Outcomes	References
ERCS focused on DNMTs: comparison with other general compound collections	Chemoinformatic analysis based on physicochemical properties, molecular fingerprints, and scaffolds	The structure of DNMTis is diverse; occupy the traditional chemical space of drugs but also cover other regions. Four tentative privileged epigenetic scaffolds were identified	Fernández-de Gortari and Medina-Franco (2015)
ERCS focused on DNMTs: comparison with epigenetic reference databases	Chemoinformatic analysis including metrics of structural complexity	The structures of DNMTis are different from inhibitors of histone deacetylases and bromodomains. There is a large potential to develop DNMTis with increased complexity	Prieto-Martínez, Fernández-de Gortari, Méndez-Lucio, Medina-Franco (2016)
Chemoinformatic-based pharmacophore model	Pharmacophore models based on the predicted protein–ligand interaction profiles of selected chemical scaffolds enriched with active compounds	A model with three main pharmacophoric points was identified. The model had high sensitivity and specificity in a validation study	This chapter
Activity landscape modeling	Systematic comparison of structure similarity and potency difference of 280 compounds using SAS maps. The landscape of two major groups of inhibitors was assessed	Inhibitors related to the cofactor have a rough landscape with significant activity cliffs. Structure-based interpretation of an activity cliffs suggested distinct molecular interactions in the binding site. Also, a novel "activity landscape sweeping" approach was proposed	Naveja and Medina-Franco (2015)

Continued

Table 1 Representative Computational Studies on DNMT Is Recently Reported—cont'd

Study	Approach	Major Outcomes	References
Linear discriminant analysis-based QSAR and virtual screening	QSAR model for 47 molecules was used to classify natural products as active/inactive; predicted active were docked with DNMT1 and 3A	Six natural products as consensus hits from two docking programs were suggested as potential inhibitors of DNMTs	Maldonado-Rojas, Olivero-Verbel, and Marrero-Ponce (2015)
Virtual screening with experimental validation	Docking-based screening of commercial compounds followed by analog searching	Identification of compound DC_05, DC_501, and DC_517 as low micromolar and selective inhibitors of DNMT1. DC_517 had activity in cell-based assays	Chen et al. (2014)
Molecular dynamics and virtual screening with experimental validation	Ensemble docking using conformations of the protein obtained from molecular dynamics. Multistep docking-based screening of a commercial compound collection	Two experimentally validated hits were identified: ASINEX ID BAS 12771472 and BAS 00872020. The former compound showed in vivo activity. A newly putative binding site in hDNMT1 is proposed	Joshi, Rajpathak, Narwade, and Deobagkar (2016)
Docking of newly designed selective compounds inspired in a virtual screening hit	Flexible docking	The selectivity of the most potent compounds toward DNMT3 over DNMT1 is rationalized at the molecular level	Aldawsari et al. (2016)
Toward computer-assisted drug repurposing	Similarity searching and data fusion	Potential DNMT inhibitors are identified	Naveja, Dueñas-González, and Medina-Franco (2016)

Table 2 Representative Confirmatory Assays in PubChem Related to DNMTs

AID	Name	Title	Assay Data	Source
1066238	Inhibition of human recombinant DNMT1 expressed in H19 cells assessed as inhibition of tritiated methyl incorporation from [3H]-labeled AdoMet into hemimethylated DNA duplex after 2 h by liquid scintillation counting analysis	Synthesis and evaluation of analogs of *N*-phthaloyl-L-tryptophan (RG108) as inhibitors of DNA methyltransferase 1	2 active, 9 tested	ChEMBL
461004	Inhibition of human recombinant DNMT3B expressed in baculovirus–insect cell system by scintillation counting	Novel and selective DNA methyltransferase inhibitors: docking-based virtual screening and experimental evaluation	2 active, 1 activity ≤1 μM, 4 tested	ChEMBL
424255	Inhibition of human recombinant DNMT1	Constrained (L-)-*S*-adenosyl-L-homocysteine (SAH) analogs as DNA methyltransferase inhibitors	18 active, 2 activity ≤1 μM, 25 tested	ChEMBL
736572	Inhibition of human DNMT1 using AdoMet and poly dI-dC after 2 h by radioactive assay	Synthetic approaches to DNMT inhibitor SGI-1027 and effects on the U937 leukemia cell line	1 active, 1 tested	ChEMBL
424222	Inhibition of human recombinant DNMT3b2 expressed in baculovirus infected high five insect cells	SAR around (L)-*S*-adenosyl-L-homocysteine, an inhibitor of human DNA methyltransferase (DNMT) enzymes	16 active, 7 activity ≤1 μM, 20 tested	ChEMBL
675178	Inhibition of DNMT1 in human HeLa cell nuclear extract assessed as methylated substrate level at 10 μM by ELISA	New cytosine derivatives as inhibitors of DNA methylation	2 tested	ChEMBL

Continued

Table 2 Representative Confirmatory Assays in PubChem Related to DNMTs—cont'd

AID	Name	Title	Assay Data	Source
657293	Inhibition of human DNMT1 using oligonucleotide 2 as substrate after 5000 s by micro plate reader based real-time break-light assay	Development of rationally designed DNA N6 adenine methyltransferase inhibitors	1 active, 1 activity $\leq$1 μM, 3 tested	ChEMBL
257909	Inhibitory activity against DNA methyl transferase in leukemic NALM6 cells	Discovery of two novel, small-molecule inhibitors of DNA methylation	2 tested	ChEMBL
631009	Inhibition of human recombinant DNMT1 expressed in Sf9 cells assessed as remaining activity at 2 mM after 3 h by scintillation counting in presence of [3H] *S*-adenosyl methionine	Synthesis and biochemical evaluation of delta(2)-isoxazoline derivatives as DNA methyltransferase 1 inhibitors	26 tested	ChEMBL
1073761	Cytotoxicity against human U937 cells after 48 h by trypan blue exclusion assay	Selective nonnucleoside inhibitors of human DNA methyltransferases active in cancer including in cancer stem cells	6 active, 6 tested	ChEMBL
602386	Dose response confirmation of DNMT1 inhibitors in a fluorescent molecular beacon assay	Dose–response confirmation of DNMT1 inhibitors in a fluorescent molecular beacon assay	179 active, 1 activity $\leq$1 μM, 200 tested	Burnham Center for Chemical Genomics
613112	Inhibition of DNMT in cell-free system	Epigenetic profiling of the antitumor natural product psammaplin A and its analogs	1 active, 1 activity $\leq$1 μM, 1 tested	ChEMBL
890	Confirmation concentration-response assay for epigenetic modulators	Confirmation concentration-response assay for epigenetic modulators	24 active, 1 activity $\leq$1 μM, 51 tested	NCGC
597	qHTS assay for epigenetic modulators	qHTS assay for epigenetic modulators	59 active, 68401 tested	NCGC

confirmatory bioassays deposited in PubChem (Wang et al., 2012). Most of these assays are also included in ChEMBL. For each assay it is reported the corresponding assay ID number (AID), name, and title as stored in PubChem, the number of tested molecules as well as the number of active and inactive compounds, and the source of the assay. SAI deposited in PubChem has been used as a starting point to conduct molecular docking and explain, at the molecular level, protein–ligand interactions (Medina-Franco & Yoo, 2013).

The growing SAI knowledge has made feasible to start charting the epigenetic relevant chemical space (ERCS) focused on DNMT1, develop new pharmacophore hypothesis, and conduct systematic structure–activity relationships (SAR) studies. Recent progress on these major topics is discussed in the next sections.

3. CHEMOINFORMATIC STUDIES OF DNMTs

The increasing amount of information available in public repositories has enabled the systematic exploration of the chemical space using chemoinformatic approaches; derive a pharmacophore model based on promising molecular scaffolds and conduct systematic SAR studies using the concept of activity landscape modeling. These studies are discussed in the sections later.

3.1 Characterization of Chemical Space: ERCS

Fernández-de Gortari et al. assembled and curated a data set with SAI of 566 nonduplicate inhibitors of DNMT1 collected from Binding Database, ChEMBL, HEMD, and literature searching recent to the time of that study (Fernández-de Gortari & Medina-Franco, 2015). The data set with 566 compounds was compared to 1490 approved drugs, a general screening collection with 1100 molecules, 837 compounds in clinical trials, and a data set of 113 molecules from a commercial library focused on epigenetic targets. The data set of inhibitors of DNMT1 was characterized using three complementary structure representations, namely; physicochemical properties of pharmaceutical interest, molecular fingerprints of different design, and molecular scaffolds using the definition of cyclic system implemented in the molecular equivalence index (MEQI) program developed by Johnson et al. (Xu & Johnson, 2002). The comparison of the databases revealed that compounds tested for DNMT1 inhibition have similar flexibility but are more polar than the reference databases (as indicated by measures of the distribution of hydrogen bond donors, acceptors, and topological surface area). Visual representation of the property space indicated that molecules tested as inhibitors of DNMT1 occupy the traditional chemical space but there are compounds that also expand the typical medicinal chemistry space. Molecular diversity analysis computed using fingerprint representations of different design confirmed the structural diversity of the compound data set. In that study, four tentative nonnucleoside privileged epigenetic scaffolds were identified, ie, molecular scaffolds with relative high frequency and high enrichment factor. As reviewed later, these selected chemical scaffolds served as starting point to develop a chemoinformatic-based pharmacophore model (Fernández-de Gortari & Medina-Franco, 2015).

Fig. 2 shows a visual representation of the chemical space of 549 compounds tested as inhibitors of DNMT1 (data points in blue) as compared to a data set of 1798 drugs obtained from Drug Bank (Law et al., 2014) (red), 1477 Generally Recognized as Safe (GRAS) compounds (yellow) (Medina-Franco, Martínez-Mayorga, Peppard, & Del Rio, 2012; Smith et al., 2011), and a data set of 2133 chemicals from the collections Everything Added to Food in the United States (EAFUS) (green) (http://www.accessdata.fda.gov/scripts/fcn/fcnnavigation.cfm?rpt=eafuslisting). The visual representation was obtained by principal component analysis of six physicochemical properties of pharmaceutical relevance commonly used to characterize compound databases, namely; molecular weight, number of rotatable bonds, number of hydrogen bond donors, number of hydrogen bond acceptors, topological surface area, and SlogP (Medina-Franco, 2012a, 2012b; Singh et al., 2009). The first three principal components capture 95.07% of the variance. Data visualization was done with the program Data Warrior (Sander, Freyss, von Korff, & Rufener, 2015). The visual representation clearly shows that compounds currently tested as DNMT1 inhibitors share the chemical space of drugs contained in Drug Bank. Compounds in Drug Bank have a large property diversity. GRAS molecules and, not surprisingly EAFUS molecules occupy an even larger fraction of the property space. Interestingly, DNMT1 compounds share the property space of GRAS and EAFUS chemicals suggesting that, at least in principle, it is possible to identify food-related chemicals with potential activity against DNMT1 (*vide infra*).

To further advance the knowledge of ERCS, Prieto-Martínez et al. compared in a follow-up work the chemical space of DNMTIs with histone deacetylases (HDACs) and bromodomains (BRDs) inhibitors (Prieto-Martínez et al., 2016). In addition of comparing the compound databases using the physicochemical properties of pharmaceutical interest, molecular

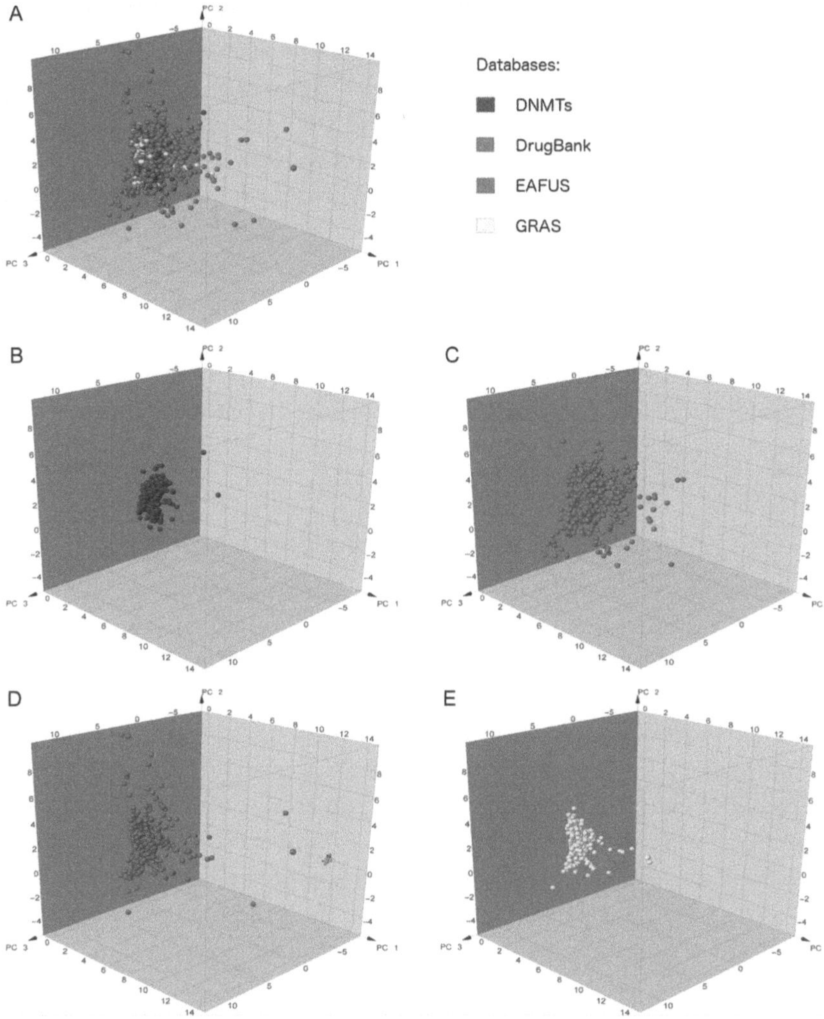

Fig. 2 Visual representation of the chemical space of 549 compounds tested as DNMT inhibitors as compared to 1798 drugs and food-related chemicals obtained by principal component analysis of six physicochemical properties of pharmaceutical relevance. Reference datasets include 1477 Generally Recognized as Safe (GRAS) compounds and a set of 2133 Everything Added to Food in the United States (EAFUS) chemicals. The first three principal components capture 95.07% of the variance.

fingerprints, and scaffolds, the structural complexity was assessed using the fraction of chiral atoms and fraction of sp^3 carbon atoms. It was concluded that the development of poly-epigenetic drugs has not been explored as initially expected. Indeed, the chemical structures of compounds tested for DNMT1 inhibition are, in general, different from the compounds tested as HDACs and BRDs inhibitors. These findings encouraged the experimental multiepigenetic target screening as an approach to evaluate the possibility to develop multiepigenetic therapies (Prieto-Martínez et al., 2016). The results of the structural complexity analysis indicated that there is a large potential to develop novel DNMT1 inhibitors with increased molecular complexity that may result in increased target selectivity (Clemons et al., 2010). In the context of polypharmacology, BRDs and HDACs inhibitors do share "common ground." This may come as no surprise as these enzymes have a complementary relationship. Additionally, the study of Prieto-Martínez et al. uncovered a significant overlap between GRAS and epi-like molecules, further reinforcing the opportunity to apply the principles of Foodinformatics to systematically identify food-related chemicals with epigenetic activity.

3.2 Chemoinformatic-Based Pharmacophore Model

Pharmacophore is defined by the IUPAC as "an ensemble of steric and electronic features that is necessary to ensure the optimal supramolecular interactions with a specific biological target and to trigger (or block) its biological response" (Wermuth, Ganellin, Lindberg, & Mitscher, 1998). Thus far, pharmacophore modeling for DNMTs has been explored on a limited basis. The first computational study was published by Yoo and Medina-Franco (2011). In that work, a structure-based pharmacophore hypothesis was derived for 14 DNMT1 inhibitors using the docked poses with therein proposed homology model of the catalytic domain of human DNMT1 (hDNMT1). Of note, at the time that work was conducted, the crystallographic structure of the catalytic domain of hDNMT1 was not available. The year after, when the first crystallographic structure of DNMT1 became available (Song, Rechkoblit, Bestor, & Patel, 2011), Yoo et al. developed an energy optimized pharmacophore (e-pharmacophore) hypothesis for 16 DNMT1 inhibitors using the same computational approach previously developed for the 14 inhibitors. However, in contrast with the previous work, the docked poses were obtained with the crystallographic structure (Yoo, Kim, Robertson, & Medina-Franco, 2012). Notably, the pharmacophore model derived with the crystal structure of DNMT1 was in good agreement with the hypothesis developed with the homology model. In a prospective application, the pharmacophore model had a good agreement with the experimental inhibitory activity of a new compound (Yoo et al., 2012).

Based on the most attractive chemical scaffolds enriched with active compounds identified in the study of Fernández-de Gortari et al. (Fernández-de Gortari & Medina-Franco, 2015), a pharmacophore model was recently developed. As discussed earlier, after systematically computing the chemical scaffolds, the most frequent scaffolds with high enrichment factors were identified. The most active compounds within the selected cyclic systems were docked with a crystallographic structure of the catalytic domain of hDNMT1 previously used in docking (Yoo et al., 2012). The Internal Coordinates Mechanics software was employed (Neves, Totrov, & Abagyan, 2012). The pocket was defined around the catalytic cysteine (Cys1226) with a volume that included the substrate and cofactor sites. Selected docked poses with the best docking scores were analyzed using protein–ligand interaction fingerprints (PLIF) with the program Molecular Operating Environment (MOE), version 2014.08 (MOE, 2014). The PLIF analysis, that represent a convenient manner to put together the chemoinformatic and molecular modeling analysis (Medina-Franco, Méndez-Lucio, & Martinez-Mayorga, 2014), was the basis to develop several pharmacophore models using the schemes "Unified" and "PPCH_All" (each scheme with different ways of annotation which may include various elements of the ligand–protein interactions). The different models were tested for their ability to identify active compounds in a data set of known DNMT1 inhibitors. The results were analyzed in terms of the ability of the models to identify false positives, false negatives, true positives, and true negatives. For each model a confusion matrix was calculated to compute the specificity and sensitivity of the hypotheses and to determine the receiving operating curve (ROC) space that, in turn, was employed to select the best query pharmacophores. The most attractive ones are those that have high sensitivity and high specificity. Of note, these measures dependent on the threshold to define an "active" compound that may vary with the objectives of the project and the experimental resources. From the ROC space it was selected the model that identified active and structural diverse (as diverse as possible) compounds of the nonnucleosidic type. One of the selected models was obtained with the PPCH_All scheme, with a maximum radius of 3 Å and feature coverage of 50%. The model was derived from the set of compounds with the core scaffold with the MEQI identifier RNDWX (Fernández-de Gortari & Medina-Franco, 2015). The selected model had three pharmacophoric points: two hydrophobic regions with average radius of 1.8 and 2.6 Å, respectively; and an H-bond acceptor, and metal ligator with average radius of 1.4 Å. The model has been tested in our group in a preliminary screening of newly designed molecules as potential inhibitors of DNMT1.

3.3 Activity Landscape Modeling

Activity landscape modeling is an approach to explore systematically the SAR of compound data sets. The method is based on systematic pairwise structure and activity comparisons of all possible combinations in a compound data set. According to the "similarity principle" it is anticipated that compounds with similar chemical structure have similar biological activities. Pairs of compounds with high similar structure but significant different biological activity are exceptions to this rule and are termed "activity cliffs" (Maggiora, 2006). Several computational methods have been proposed to conduct the pairwise structure–activity comparisons of data sets using the concept of activity landscape. One of such methods is the structure–activity similarity (SAS) maps (and related plots) (Medina-Franco, 2012a, 2012b; Medina-Franco & Waddell, 2012). Activity landscape methods have emerged as valuable tools to understand the SAR before attempt making predictions using classical approaches such as quantitative structure–activity relationships (QSAR) (Medina-Franco, Navarrete-Vázquez, & Méndez-Lucio, 2015).

Naveja et al. conducted an activity landscape study of 280 compounds retrieved from ChEMBL, version 20 (Naveja & Medina-Franco, 2015). Exploratory analysis of the chemical space of the data set clearly showed the presence of two types of molecules in the data set: SAM-analogs and nonnucleoside compounds. That finding led to the independent exploration of the activity landscape of the two types of compounds. In other words, it was assessed the "local" activity landscape of SAM-analogs and nonnucleoside molecules. The structure similarity was measured with the Tanimoto coefficient and two different structural fingerprints (along with the combination of the similarity values computed with both). For each pair of compounds, the relationship between the structure and the activity similarity was portrayed using SAS maps. It was concluded that compounds structurally related to the cofactor have activity cliffs with significant differences in the biological activities. In contrast, nonnucleoside molecules have fewer activity cliffs and the activity landscape is, overall, smoother as compared to the landscape of the molecules related to the cofactor. One of the practical applications of the findings was that essentially any compound not related to the cofactor can be used as a query (reference) molecule in similarity searching. A second major conclusion was that, at least in principle, the nonnucleosidic molecules contained in ChEMBL (version 20) can be the used as basis to develop QSAR models (Naveja & Medina-Franco, 2015).

As part of the findings, compound CHEMBL3126646, that is a regioisomer of SGI-1027 (Fig. 1), emerged as an "activity cliff generator" defined by Mendez-Lucio et al. as a molecule with high propensity to form activity cliffs (Mendez-Lucio, Perez-Villanueva, Castillo, & Medina-Franco, 2012). Structure-based analysis of the activity cliff generator was in agreement with the experimental information indicating that CHEMBL3126646 is not a competitive inhibitor of the cofactor. Furthermore, docking with DNMT1 suggested that the carbonyl oxygen of CHEMBL3126646 may form a hydrogen bond with the O_2' oxygen atom of *S*-adenosyl-L-homocysteine (SAH) (Naveja & Medina-Franco, 2015). However, the precise binding site of these and several other DNMTis remains an open question.

3.4 Quantitative Structure–Activity Relationships

Maldonado-Rojas et al. recently reported a linear discriminant analysis-based QSAR model for 47 compounds obtained from PubChem with known biological activity against DNMTs. Out of the total molecules, 32 were included in the training set and 15 in the test set. The model was developed to classify compounds as active or inactive and the resultant model had an accuracy of 93%. In a prospective application, the model was employed to predict the classification of 800 natural products as active/inactive. The predicted active molecules, 447 in total, were subject to molecular docking with crystallographic structures of DNMT1 and DNMT3A using two docking programs: AutoDock Vina and Surflex-Dock. Six consensus hits were selected based on three criteria, namely; score values, protein–ligand interactions, and structural diversity. The six natural products were 9,10-dihydro-12-hydroxygambogic, phloridzin, 2′,4′-dihydroxychalcone 4′-glucoside, daunorubicin, pyrromycin, and centaurein. It remains to conduct the experimental validation of the computational hits (Maldonado-Rojas et al., 2015). A docking-based virtual screening (VS) of natural products as potential inhibitors of DNMTs have been previously reported (Medina-Franco, López-Vallejo, Kuck, & Lyko, 2011). These studies are further examples of the synergy between natural product-based drug discovery with computational methods (Do, Medina-Franco, Scior, & Bernard, 2015; Medina-Franco, 2013).

4. VS: HIT IDENTIFICATION AND OPTIMIZATION

VS has been a successful approach to filter large compound databases and identify novel hits compounds that, in turn, have been starting points to optimize the activity. Examples of VS hits are RG108, NSC14778 (Fig. 3) (Kuck, Singh, Lyko, & Medina-Franco, 2010; Siedlecki et al., 2006) which subsequent optimization had been reviewed (Medina-Franco, Méndez-Lucio, et al., 2015). Compound collections that have been screened include public collections, natural products, and commercial libraries, to name a few. Structure-based VS continue to be a source of hit compounds as described in this section. Fig. 3 outlines the continued progress on the optimization and development of experimentally validated VS hits, many of which have been performed with the aid of computational approaches.

4.1 Novel VS Hits

Chen et al. recently reported the identification of compound DC_05 (Fig. 3), a low micromolar ($IC_{50}=10.3$ μM) and selective nonnucleoside inhibitor of DNMT1 (Chen et al., 2014). Authors conducted a docking-based VS of a commercial screening collection with 111,121 compounds (considering the removal of molecules with unfavorable physicochemical properties) followed by experimental validation. The docking was conducted with the program Glide in the catalytic and cofactor sites of a crystal structure of mouse DNMT1 bound to SAH. After docking, 51 structurally diverse compounds were acquired and tested in a biochemical assay. DC_05 showed activity with DNMT1 and selectivity toward other AdoMet-dependant protein methyltransferases, including DNMT3A and DNMT3B.

In order to explore the local SAR and identify more potent compounds than DC_05, the confirmed hit identified from docking was used as query molecule to conduct analog searching. A total of 19 molecules with similar scaffolds to DC_05 were purchased and tested; 8 molecules were identified as inhibitors of DNMT1 at the micromolar level and 2 compounds, DC_501 ($IC_{50}=2.5$ μM) and DC_517 ($IC_{50}=1.7$ μM) (Fig. 3) showed better activity than the parent molecule. Analysis of the biological activities of the enantiomers of DC_05 and DC_517 were compared to their racemates showing similar potencies (Chen et al., 2014). In cell-based assays, DC-517 showed the most promising antiproliferative effects in HCT116 (human colon cancer) and Capan-1 (human pancreatic adenocarcinoma cells) (Chen et al., 2014).

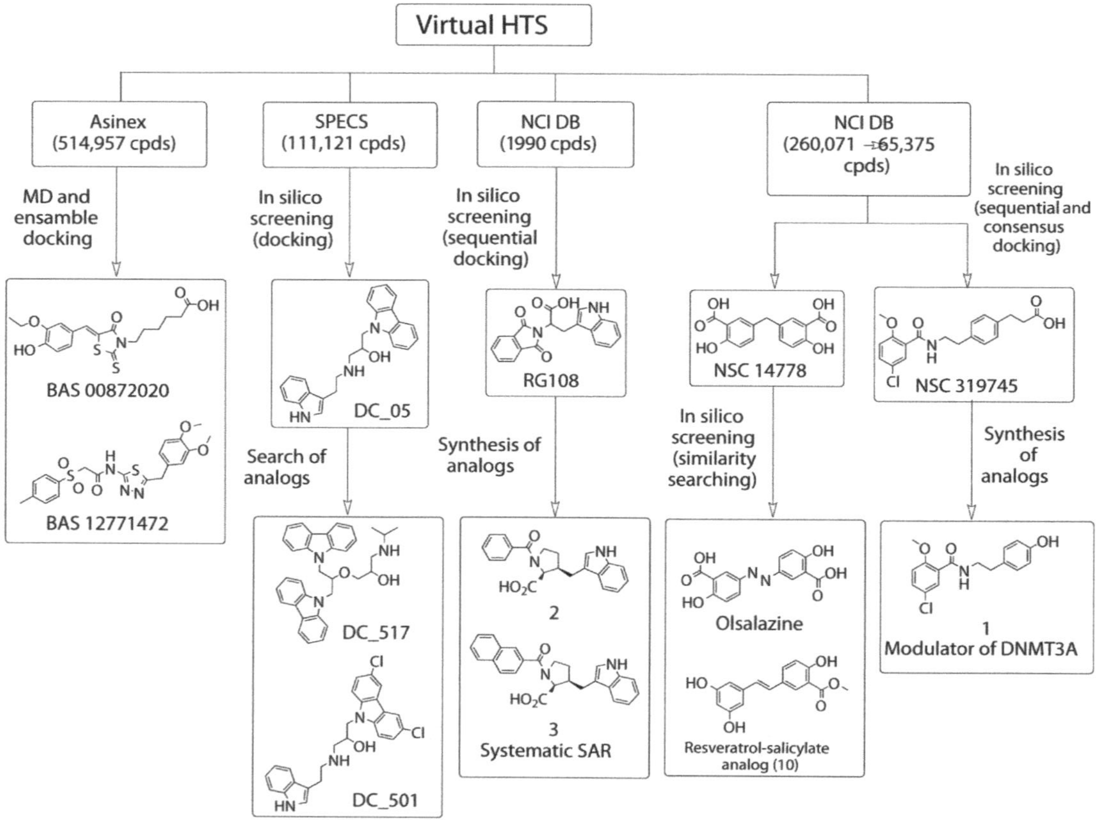

Fig. 3 Computational-driven identification and development of DNMTis and molecules with hypomethylating activities. The figure represents and update of the synergy between virtual high-throughput screening (HTS) and medicinal chemistry (Medina-Franco, Méndez-Lucio, et al., 2015).

In a separate study, Joshi et al. conducted several molecular dynamics simulations at the nanosecond time scale of the active form of hDNMT1. The output of the molecular dynamics was used as structural information to conduct an ensemble docking-based VS of a commercial screening collection (Asinex) with 514,957 compounds. The rationale to select this database was to screen compound collections different from the previously used in VS (Kuck et al., 2010; Medina-Franco & Yoo, 2013; Siedlecki et al., 2006). The docking was conducted implementing a multistep docking approach with the program Glide using a more precise scoring function at each step. The docking search was focused on the cavity of DNMT1 that encompassed the cofactor SAM, cytosine, and the 5-mC binding sites. Eight compounds were selected as computational hits and experimentally tested. Two small molecules ASINEX ID BAS 12771472 (compound A in the original paper) and BAS 00872020 were found to inhibit hDNMT1 in in vitro assays at the micromolar level with IC_{50} values of 15 and 18 μM, respectively (Fig. 3). In order to experimentally elucidate the mechanism of action of the inhibitors, competition assays with SAM were conducted. Results showed that both hit molecules do not compete with the cofactor. Compound "A" also exhibited in vivo activity by reducing the levels of total genomic methylation (Joshi et al., 2016). The authors of that work also concluded that the structure information provided by the molecular dynamics simulations represent a novel unique site to identify selective inhibitors toward hDNMT1 (Joshi et al., 2016).

4.2 Follow-Up of VS Hits

As reviewed earlier, several VS hits have been used as starting points of optimization programs (Medina-Franco, Méndez-Lucio, et al., 2015). One reviewed example is a novel modulator of DNMT3A (1) that was developed based on the chemical structure of NSC319745 (Fig. 3) (Kabro et al., 2013). The later compound was identified from the computational screening of the NCI database (Kuck et al., 2010).

A second experimentally validated hit of the VS of the NCI database was methylenedisalicylic acid, NSC14778 (Fig. 3) (*vide supra*). In a recent work by Aldawsari et al. the chemical structure of the methylenedisalicylic acid provided further rationale to synthesize and screen resveratrol-salicylate analogs. Some of the newly designed compounds showed selective inhibition against DNMT3 enzymes. One of the most active compounds, as measured under the assay conditions of that work, was compound "**10**" (as numbered in the original paper) (Fig. 3) that showed IC_{50}'s equal to 25 and 62 μM against DNMT3A and DNMT3B, respectively. This molecule also showed remarkable cytotoxicity against three human cancer cells (Aldawsari et al., 2016). Molecular docking of the most active compounds was performed to rationalize the interactions with DNMT3A and DNMT3B at the molecular level. The docking was performed with the program AutoDock. Since it is uncertain the precise binding pocket of the resveratrol-salicylate analogs, the docking studies were performed in the presence and absence of the cofactor. One of the most relevant findings was that the predicted interactions with DNMTs were able to rationalize the observed enzymatic inhibitory selectivity. Interestingly, despite the well-known limitation of docking algorithms to accurately calculate the binding energies, the calculated binding energies with DNMTs were more favored (less negative) than the calculated binding energies with DNMT1. This result was in excellent agreement with the data of the experimental evaluation (Aldawsari et al., 2016).

5. COMPUTER-ASSISTED DRUG REPURPOSING

Drug repurposing (also called repositioning) aims to identify a novel medical indication for a drug. The drug can currently be approved for an alternative clinical use, withdrawn from the market, for example, because of side effects or because of lack of efficacy (Ashburn & Thor, 2004; Würth et al., 2016). In light of the increasing awareness of polypharmacology, that is a concept highly attached to drug repurposing (Medina-Franco, Giulianotti, Welmaker, & Houghten, 2013), there is an interest to identify approved drugs for other indications as hypomethylating agents. In fact, there are reported drugs such as procaine, procainamide, and hydralazine, to name a few, that have been described as hypomethylating compounds (Medina-Franco, Méndez-Lucio, et al., 2015). However, the hypomethylating characteristics have been identified fortuitously and there is a current effort to conduct systematic drug repurposing for DNMTis and other epigenetic targets (Naveja et al., 2016). Computational approaches, in conjunction with experimental approaches, offer an avenue to conduct systematic drug repurposing.

One of the first reported efforts toward the computer-aided drug repurposing for DNMT1 inhibitors is the work of Méndez-Lucio et al. that identified olsalazine (an antiinflammatory drug), as a novel hypomethylating agent (Méndez-Lucio, Tran, Medina-Franco, Meurice, & Muller, 2014).

In a recent follow-up work, Naveja et al. implemented a similarity-based VS approach to identify in Drug Bank, a public molecular database of approved drugs (Law et al., 2014), small molecules with high structural similarity to molecules with known inhibitory activity against DNMT1 (Naveja et al., 2016). The working hypothesis of similarity searching is that "similar compounds are expected to have similar activities" with the exception of activity cliffs (*vide supra*). In that study, a set of 13 structural diverse and active compounds with DNMT1 were selected from ChEMBL as reference (query) molecules. The queries were used to measure the maximum and mean structural similarity of each compound in Drug Bank with the 13 reference compounds using the Tanimoto coefficient and two fingerprint representations of different design. The approved drugs in Drug Bank most similar to the 13 reference DNMT1 inhibitors were selected using multifusion similarity plots (Medina-Franco, Maggiora, Giulianotti, Pinilla, & Houghten, 2007) by choosing the compounds with the overall highest maximum and mean similarity values. The next logical step of that work is to conduct the experimental validation of the VS hits (Naveja et al., 2016).

6. FOOD CHEMICALS AS POTENTIAL MODULATORS OF DNMTs AND OTHER EPIGENETIC TARGETS

A growing body of evidence suggests that nutrients and environmental exposures play a major role in establishing abnormal DNA methylation patterns and other epigenetic mechanisms. Therefore, a constant uptake of DNA demethylating agents is believed to have a chemopreventive effect. In principle, this could be conveniently achieved through the dietary uptake of food chemicals (or other "safe" natural products) that modulates DNMT. A prominent example of this hypothesis is the polyphenol compound from green tea, (−)-epigallocathechin-3-gallate (EGCG), which has been proposed to inhibit DNMT1 and reactivate methylation-silenced genes in cancer cells. Systematic mining of food chemicals as potential modulators of DNMTs or, more general, epigenetic targets is a challenging task since food is a multivariable system. As discussed by Martinez-Mayorga and Peña Montes, composition of raw food may be different through the year and may vary by geographical region. In addition, portion sizes and cooking habits have an influence on human health (Martinez-Mayorga & Montes, 2016). Despite these challenges, it has been of interest the systematic identification of dietary components as modulators of the epigenome. The same molecular modeling approaches used to mine and conduct VS of screening

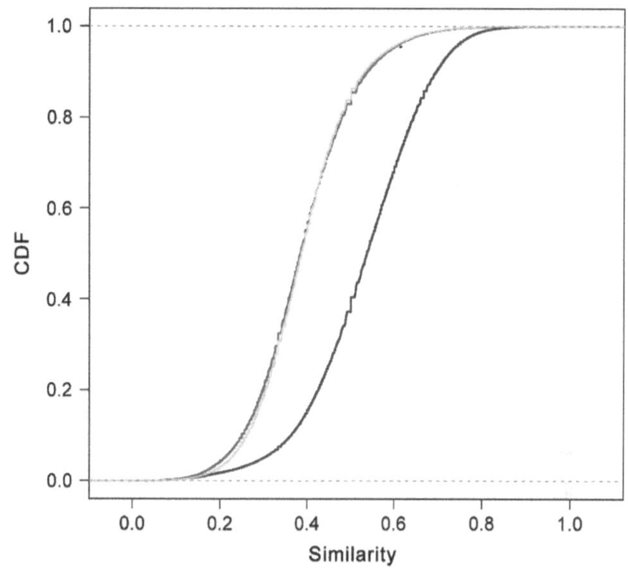

Fig. 4 Nearest-neighbor curves of the structure similarity of approved drugs (*blue* (*black* in the print version)), GRAS (*green* (*light gray* in the print version)), and EAFUS (*red* (*gray* in the print version)) compounds to compounds tested as inhibitors of DNMT1. The structure similarity was computed with MACCS keys (166-bits) and the Tanimoto coefficient.

collections can be implemented to interrogate the chemical space of food chemicals (Ruddigkeit & Reymond, 2014). For example, Fig. 4 shows the nearest-neighbor curves, ie, distribution of the maximum similarity values, of approved drugs (blue curve), GRAS (green), and EAFUS (red) compounds to compounds tested as inhibitors of DNMT1 (used as reference molecules). The similarity values were computed with MACCS keys (166-bits) and the Tanimoto coefficient. The curved in Fig. 4 clearly shows that, approved drugs are, in general, more similar to DNMT1 compounds than GRAS and EAFUS chemicals. In turn, the food-related databases have comparable profiles of structure similarity to DNMT1 molecules. Interestingly, there are GRAS molecules with identical structure representation as DNMT1 molecules.

7. CONCLUDING REMARKS

The role of DNMTs in epigenetic-related disorders was initially associated with cancer therapy but currently the potential use is expanding to other areas including neurodegenerative diseases and diabetes. Despite the fact additional research needs to be conducted to fully understand the relationship between DNMTs and other epigenetic targets, the development of nonnucleoside inhibitors of DNMT as epi-drugs and epi-probes continue to be an intense area of research. This has been reflected in the growing SAI that is public available. This information has boosted the exploration of SAR using quantitative approaches such as activity landscape modeling and QSAR. Combining these two approaches with molecular docking has led to suggest novel mechanisms of inhibition of potent inhibitors such as CHEMBL3126646 and select VS hits for experimental evaluation. Computational screening of compound databases, followed by experimental validation continue to uncover DNMTis with novel chemical scaffolds that represent starting points of medicinal chemistry optimization programs. Molecular dynamics combined with ensemble docking and biochemical characterization of new DNMTis gave rise to suggest novel binding sites to design or screen selective compounds. Recent applications of docking include the explanation at the molecular level of the observed selectivity for small molecules in enzymatic inhibition assays. It is anticipated that the experimental screening of focused libraries and newly designed molecules will continue to increase the understanding of the activity of these compounds at the molecular level. It also continues the quest to elucidate the precise binding site of small-molecule DNMTis as this is one of the major challenges that face the structure-based optimization of hit and lead compounds. Thus far, docking-based VS has been successful in identifying hit compound with novel chemical scaffolds suggesting that the hypothesis of the putative binding site is valid. However, the eventual availability of a crystallographic structure or convincing biophysical data will shed light into the precise binding position of the compounds. Computational methods in synergy with biophysical methods would play a major role for this endeavor. It is also expected that the same or similar methods commonly employed to identify potential drug candidates be used to uncover food-related chemicals with health-promoting benefits.

ACKNOWLEDGMENTS

F.P. and E.F. thank CONACyT for the fellowships No. 660465/576637 and 348291/240072, respectively. This work was supported by the Universidad Nacional Autónoma de México (UNAM), grant *Programa de Apoyo a Proyectos de Investigación e Innovación Tecnológica* (PAPIIT) IA204016. We also thank the institutional program *Nuevas Alternativas de Tratamiento para Enfermedades Infecciosas* (NUATEI) of the Instituto de Investigaciones Biomédicas (IIB) UNAM for financial support.

REFERENCES

Aldawsari, F. S., Aguayo-Ortiz, R., Kapilashrami, K., Yoo, J., Luo, M., Medina-Franco, J. L., et al. (2016). Resveratrol-salicylate derivatives as selective DNMT3 inhibitors and anticancer agents. *Journal of Enzyme Inhibition and Medicinal Chemistry*, in press. http://dx.doi.org/10.3109/14756366.2015.1058256.

Arguelles, A. O., Meruvu, S., Bowman, J. D., & Choudhury, M. (2016). Are epigenetic drugs for diabetes and obesity at our door step? *Drug Discovery Today*, *21*, 499–509. http://dx.doi.org/10.1016/j.drudis.2015.12.001.

Arrowsmith, C. H., Audia, J. E., Austin, C., Baell, J., Bennett, J., Blagg, J., et al. (2015). The promise and peril of chemical probes. *Nature Chemical Biology*, *11*(8), 536–541. http://dx.doi.org/10.1038/nchembio.1867.

Ashburn, T. T., & Thor, K. B. (2004). Drug repositioning: Identifying and developing new uses for existing drugs. *Nature Reviews. Drug Discovery*, *3*(8), 673–683. http://dx.doi.org/10.1038/nrd1468.

Chen, S. J., Wang, Y. L., Zhou, W., Li, S. S., Peng, J. L., Shi, Z., et al. (2014). Identifying novel selective non-nucleoside DNA methyltransferase 1 inhibitors through docking-based virtual screening. *Journal of Medicinal Chemistry*, *57*(21), 9028–9041. http://dx.doi.org/10.1021/jm501134e.

Clemons, P. A., Bodycombe, N. E., Carrinski, H. A., Wilson, J. A., Shamji, A. F., Wagner, B. K., et al. (2010). Small molecules of different origins have distinct distributions of structural complexity that correlate with protein-binding profiles. *Proceedings of the National Academy of Sciences of the United States of America*, *107*(44), 18787–18792. http://dx.doi.org/10.1073/pnas.1012741107.

Derissen, E. J. B., Beijnen, J. H., & Schellens, J. H. M. (2013). Concise drug review: Azacitidine and decitabine. *The Oncologist*, *18*(5), 619–624. http://dx.doi.org/10.1634/theoncologist.2012-0465.

Do, Q. T., Medina-Franco, J. L., Scior, T., & Bernard, P. (2015). How to valorize biodiversity? Let's go hashing, extracting, filtering, mining, fishing. *Planta Medica*, *81*, 436–449. http://dx.doi.org/10.1055/s-0034-1396314.

Erdmann, A., Arimondo, P. B., & Guianvarc'h, D. (2016). Structure-guided optimization of DNA methyltransferase inhibitors. In J. L. Medina-Franco (Ed.), *Epi-informatics* (pp. 53–74). London, UK: Academic Press.

Erdmann, A., Halby, L., Fahy, J., & Arimondo, P. B. (2014). Targeting DNA methylation with small molecules: What's next? *Journal of Medicinal Chemistry*, *58*, 2569–2583. http://dx.doi.org/10.1021/jm500843d.

Fernández-de Gortari, E., & Medina-Franco, J. L. (2015). Epigenetic relevant chemical space: A chemoinformatic characterization of inhibitors of DNA methyltransferases. *RSC Advances*, *5*(106), 87465–87476. http://dx.doi.org/10.1039/C5RA19611F.

Gaulton, A., Bellis, L. J., Bento, A. P., Chambers, J., Davies, M., Hersey, A., et al. (2012). ChEMBL: A large-scale bioactivity database for drug discovery. *Nucleic Acids Research*, *40*(D1), D1100–D1107. http://dx.doi.org/10.1093/nar/gkr777.

Gros, C., Fahy, J., Halby, L., Dufau, I., Erdmann, A., Gregoire, J.-M., et al. (2012). DNA methylation inhibitors in cancer: Recent and future approaches. *Biochimie*, *94*(11), 2280–2296. http://dx.doi.org/10.1016/j.biochi.2012.07.025.

Guianvarc'h, D., & Arimondo, P. B. (2014). Challenges in developing novel DNA methyltransferases inhibitors for cancer therapy. *Future Medicinal Chemistry*, *6*(11), 1237–1240. http://dx.doi.org/10.4155/fmc.14.82.

Huang, Z., Jiang, H., Liu, X., Chen, Y., Wong, J., Wang, Q., et al. (2012). HEMD: An integrated tool of human epigenetic enzymes and chemical modulators for therapeutics. *Plos One*, 7(6), e39917. http://dx.doi.org/10.1371/journal.pone.0039917.

Joshi, M., Rajpathak, S. N., Narwade, S. C., & Deobagkar, D. (2016). Ensemble based virtual screening and experimental validation of inhibitors targeting a novel site of human DNMT1. *Chemical Biology and Drug Design*, in press. http://dx.doi.org/10.1111/cbdd.12741.

Kabro, A., Lachance, H., Marcoux-Archambault, I., Perrier, V., Dore, V., Gros, C., et al. (2013). Preparation of phenylethylbenzamide derivatives as modulators of DNMT3 activity. *Medicinal Chemistry Communications*, *4*(12), 1562–1570. http://dx.doi.org/10.1039/c3md00214d.

Kuck, D., Singh, N., Lyko, F., & Medina-Franco, J. L. (2010). Novel and selective DNA methyltransferase inhibitors: Docking-based virtual screening and experimental evaluation. *Bioorganic and Medicinal Chemistry*, *18*(2), 822–829. http://dx.doi.org/10.1016/j.bmc.2009.11.050.

Law, V., Knox, C., Djoumbou, Y., Jewison, T., Guo, A. C., Liu, Y., et al. (2014). DrugBank 4.0: Shedding new light on drug metabolism. *Nucleic Acids Research*, *42*(D1), D1091–D1097. http://dx.doi.org/10.1093/nar/gkt1068.

Liu, T. Q., Lin, Y. M., Wen, X., Jorissen, R. N., & Gilson, M. K. (2007). BindingDB: A web-accessible database of experimentally determined protein-ligand binding affinities. *Nucleic Acids Research*, *35*, D198–D201. http://dx.doi.org/10.1093/nar/gkl999.

Loharch, S., Bhutani, I., Jain, K., Gupta, P., Sahoo, D. K., & Parkesh, R. (2015). EpiDBase: A manually curated database for small molecule modulators of epigenetic landscape. *Database*, 2015, bav013. http://dx.doi.org/10.1093/database/bav013.

Maggiora, G. M. (2006). On outliers and activity cliffs-why QSAR often disappoints. *Journal of Chemical Information and Modeling*, *46*(4), 1535. http://dx.doi.org/10.1021/ci060117s.

Maldonado-Rojas, W., Olivero-Verbel, J., & Marrero-Ponce, Y. (2015). Computational fishing of new DNA methyltransferase inhibitors from natural products. *Journal of Molecular Graphics and Modelling*, *60*, 43–54. http://dx.doi.org/10.1016/j.jmgm.2015.04.010.

Martinez-Mayorga, K., & Montes, C. P. (2016). The role of nutrition in epigenetics and recent advances of in silico studies. In J. L. Medina-Franco (Ed.), *Epi-informatics* (pp. 385–398). London, UK: Academic Press. http://dx.doi.org/10.1016/B978-0-12-802808-7.00003-4.

Medina-Franco, J. L. (2012a). Interrogating novel areas of chemical space for drug discovery using chemoinformatics. *Drug Development Research*, *73*(7), 430–438. http://dx.doi.org/10.1002/ddr.21034.

Medina-Franco, J. L. (2012b). Scanning structure–activity relationships with structure–activity similarity and related maps: From consensus activity cliffs to selectivity switches. *Journal of Chemical Information and Modeling*, *52*(10), 2485–2493. http://dx.doi.org/10.1021/ci300362x.

Medina-Franco, J. L. (2013). Advances in computational approaches for drug discovery based on natural products. *Revista Latinoamericana de Quimica*, *41*(2), 95–110.

Medina-Franco, J. L. (Ed.). (2016). *Epi-informatics: Discovery and development of small molecule epigenetic drugs and probe*. London, UK: Academic Press. http://dx.doi.org/10.1016/B978-0-12-802808-7.01001-7.

Medina-Franco, J. L., Giulianotti, M. A., Welmaker, G. S., & Houghten, R. A. (2013). Shifting from the single to the multitarget paradigm in drug discovery. *Drug Discovery Today*, *18*(9–10), 495–501. http://dx.doi.org/10.1016/j.drudis.2013.01.008.

Medina-Franco, J. L., López-Vallejo, F., Kuck, D., & Lyko, F. (2011). Natural products as DNA methyltransferase inhibitors: A computer-aided discovery approach. *Molecular Diversity*, *15*(2), 293–304. http://dx.doi.org/10.1007/s11030-010-9262-5.

Medina-Franco, J. L., Maggiora, G. M., Giulianotti, M. A., Pinilla, C., & Houghten, R. A. (2007). A similarity-based data-fusion approach to the visual characterization and comparison of compound databases. *Chemical Biology and Drug Design*, *70*(5), 393–412. http://dx.doi.org/10.1111/j.1747-0285.2007.00579.x.

Medina-Franco, J. L., Martínez-Mayorga, K., Peppard, T. L., & Del Rio, A. (2012). Chemoinformatic analysis of GRAS (Generally Recognized as Safe) flavor chemicals and natural products. *PloS One*, 7(11), e50798. http://dx.doi.org/10.1371/journal.pone.0050798.

Medina-Franco, J. L., Méndez-Lucio, O., & Martinez-Mayorga, K. (2014). Chapter One—The interplay between molecular modeling and chemoinformatics to characterize protein–ligand and protein–protein interactions landscapes for drug discovery. *Advances in Protein Chemistry and Structural Biology, 96*, 1–37. http://dx.doi.org/10.1016/bs.apcsb.2014.06.001.

Medina-Franco, J. L., Méndez-Lucio, O., Yoo, J., & Dueñas, A. (2015). Discovery and development of DNA methyltransferase inhibitors using in silico approaches. *Drug Discovery Today, 20*(6), 569–577. http://dx.doi.org/10.1016/j.drudis.2014.12.007.

Medina-Franco, J. L., Navarrete-Vázquez, G., & Méndez-Lucio, O. (2015). Property landscape modeling is at the interface of chemoinformatics and experimental sciences. *Future Medicinal Chemistry*, 7(9), 1197–1211. http://dx.doi.org/10.4155/fmc.15.51.

Medina-Franco, J. L., & Waddell, J. (2012). Towards the bioassay activity landscape modeling in compound databases. *Journal of the Mexican Chemical Society, 56*(2), 163–168.

Medina-Franco, J. L., & Yoo, J. (2013). Docking of a novel DNA methyltransferase inhibitor identified from high-throughput screening: Insights to unveil inhibitors in chemical databases. *Molecular Diversity, 17*(2), 337–344. http://dx.doi.org/10.1007/s11030-013-9428-z.

Medina-Franco, J. L., Yoo, J., & Dueñas-Gonzalez, A. (2015). *DNA methyltransferase inhibitors for cancer therapy. In Y. G. Zheng (Ed.), Epigenetic technological applications* (pp. 265–290). London: Elsevier. http://dx.doi.org/10.1016/B978-0-12-801080-8.00013-2.

Mendez-Lucio, O., Perez-Villanueva, J., Castillo, R., & Medina-Franco, J. L. (2012). Identifying activity cliff generators of PPAR ligands using SAS maps. *Molecular Informatics, 31*(11–12), 837–846. http://dx.doi.org/10.1002/minf.201200078.

Méndez-Lucio, O., Tran, J., Medina-Franco, J. L., Meurice, N., & Muller, M. (2014). Towards drug repurposing in epigenetics: Olsalazine as a novel hypomethylating compound active in a cellular context. *ChemMedChem, 9*, 560–565. http://dx.doi.org/10.1002/cmdc.201300555.

Molecular Operating Environment (MOE), version 2014.08, Chemical Computing Group Inc., Montreal, Quebec, Canada. Available at http://www.chemcomp.com.

Naveja, J. J., Dueñas-González, A., & Medina-Franco, J. L. (2016). Drug repurposing for epigenetic targets guided by computational methods. In J. L. Medina-Franco (Ed.), *Epi-informatics* (pp. 327–358). London: Academic Press. http://dx.doi.org/10.1016/B978-0-12-802808-7.00012-5.

Naveja, J. J., & Medina-Franco, J. L. (2015). Activity landscape sweeping: Insights into the mechanism of inhibition and optimization of DNMT1 inhibitors. *RSC Advances, 5*(78), 63882–63895. http://dx.doi.org/10.1039/C5RA12339A.

Neves, M. A. C., Totrov, M., & Abagyan, R. (2012). Docking and scoring with ICM: The benchmarking results and strategies for improvement. *Journal of Computer-Aided Molecular Design, 26*(6), 675–686. http://dx.doi.org/10.1007/s10822-012-9547-0.

Prieto-Martínez, F. D., Fernández-de Gortari, E., Méndez-Lucio, O., & Medina-Franco, J. L. (2016). An epigenetic relevant chemical space Odyssey, *RSC Advances*, accepted.

Robertson, K. D. (2001). DNA methylation, methyltransferases, and cancer. *Oncogene, 20*(24), 3139–3155. http://dx.doi.org/10.1038/sj.onc.1204341.

Ruddigkeit, L., & Reymond, J.-L. (2014). The chemical space of flavours. In K. Martinez-Mayorga & J. L. Medina-Franco (Eds.), *FoodInformatics: Applications of chemical information to food chemistry* (pp. 83–96). Heidelberg: Springer.

Sander, T., Freyss, J., von Korff, M., & Rufener, C. (2015). DataWarrior: An open-source program for chemistry aware data visualization and analysis. *Journal of Chemical Information and Modeling, 55*(2), 460–473. http://dx.doi.org/10.1021/ci500588j.

Siedlecki, P., Boy, R. G., Musch, T., Brueckner, B., Suhai, S., Lyko, F., et al. (2006). Discovery of two novel, small-molecule inhibitors of DNA methylation. *Journal of Medicinal Chemistry, 49*(2), 678–683. http://dx.doi.org/10.1021/jm050844z.

Singh, N., Guha, R., Giulianotti, M. A., Pinilla, C., Houghten, R. A., & Medina-Franco, J. L. (2009). Chemoinformatic analysis of combinatorial libraries, drugs, natural products, and molecular libraries small molecule repository. *Journal of Chemical Information and Modeling*, *49*(4), 1010–1024. http://dx.doi.org/10.1021/ci800426u.

Smith, R. L., Waddell, W. J., Cohen, S. M., Fukushima, S., Gooderham, N. J., Hecht, S. S., et al. (2011). GRAS flavoring substances 25. *Food Technology*, *65*(7), 44–55.

Song, J., Rechkoblit, O., Bestor, T. H., & Patel, D. J. (2011). Structure of DNMT1–DNA complex reveals a role for autoinhibition in maintenance DNA methylation. *Science*, *331*(6020), 1036–1040. http://dx.doi.org/10.1126/science.1195380.

Wang, Y., Xiao, J., Suzek, T. O., Zhang, J., Wang, J., Zhou, Z., et al. (2012). PubChem's BioAssay database. *Nucleic Acids Research*, *40*(D1), D400–D412. http://dx.doi.org/10.1093/nar/gkr1132.

Wermuth, C. G., Ganellin, C. R., Lindberg, P., & Mitscher, L. A. (1998). Glossary of terms used in medicinal chemistry (IUPAC Recommendations 1998). *Pure and Applied Chemistry*, *70*, 1129. http://dx.doi.org/10.1351/pac199870051129.

Würth, R., Thellung, S., Bajetto, A., Mazzanti, M., Florio, T., & Barbieri, F. (2016). Drug-repositioning opportunities for cancer therapy: Novel molecular targets for known compounds. *Drug Discovery Today*, *21*(1), 190–199. http://dx.doi.org/10.1016/j.drudis.2015.09.017.

Xu, Y. J., & Johnson, M. (2002). Using molecular equivalence numbers to visually explore structural features that distinguish chemical libraries. *Journal of Chemical Information and Computer Science*, *42*(4), 912–926. http://dx.doi.org/10.1021/ci025535l.

Yoo, J., Kim, J. H., Robertson, K. D., & Medina-Franco, J. L. (2012). Molecular modeling of inhibitors of human DNA methyltransferase with a crystal structure: Discovery of a novel DNMT1 inhibitor. *Advances in Protein Chemistry and Structure Biology*, *87*, 219–247. http://dx.doi.org/10.1016/B978-0-12-398312-1.00008-1.

Yoo, J., & Medina-Franco, J. L. (2011). Homology modeling, docking, and structure-based pharmacophore of inhibitors of DNA methyltransferase. *Journal of Computer-Aided Molecular Design*, *25*(6), 555–567. http://dx.doi.org/10.1007/s10822-011-9441-1.

Yoo, J., & Medina-Franco, J. L. (2012). Inhibitors of DNA methyltransferases: Insights from computational studies. *Current Medicinal Chemistry*, *19*(21), 3475–3487. http://dx.doi.org/10.2174/092986712801323289.

CHAPTER TWO

Recent Advances in Computational Models for the Study of Protein–Peptide Interactions

D. Kilburg*,†, E. Gallicchio*,†,1

*Brooklyn College, Brooklyn, NY, United States

†The Graduate Center of the City University of New York, New York, NY, United States

[1]Corresponding author: e-mail address: egallicchio@brooklyn.cuny.edu

Contents

1. Introduction 28
2. Theory and Methods 30
   2.1 Statistical Mechanics Formulation of Molecular Association 30
   2.2 Pathway Free Energy Methods 32
   2.3 MM-PBSA, MM-GBSA 35
   2.4 Software for Protein–Peptide Docking 36
3. Sample Applications 39
   3.1 Predicting Protein–Peptide Interactions with Docking 39
   3.2 Conformational and Energetic Analysis by Molecular Dynamics 41
   3.3 Modeling Key Interactions with Mutagenic Analysis to Improve Affinity and Selectivity of Binding 43
   3.4 Distinguishing Binders from Nonbinders 45
   3.5 Protein–Peptide Binding Affinities 47
4. Concluding Remarks 51

References 51

Abstract

We review computational models and software tools in current use for the study of protein–peptide interactions. Peptides and peptide derivatives are growing in interest as therapeutic agents to target protein–protein interactions. Protein–protein interactions are pervasive in biological systems and are responsible for the regulation of critical functions within the cell. Mutations or dysregulation of expression can alter the network of interactions among proteins and cause diseases such as cancer. Protein–protein binding interfaces, which are often large, shallow, and relatively feature-less, are difficult to target with small-molecule inhibitors. Peptide derivatives based on the binding motifs present in the target protein complex are increasingly drawing interest as superior alternatives to conventional small-molecule inhibitors. However, the design of

Advances in Protein Chemistry and Structural Biology, Volume 105
ISSN 1876-1623
http://dx.doi.org/10.1016/bs.apcsb.2016.06.002

peptide-based inhibitors also presents novel challenges. Peptides are more complex and more flexible than standard medicinal compounds. They also tend to form more extended and more complex interactions with their protein targets. Computational modeling is increasingly being employed to supplement synthetic and biochemical work to offer guidance and energetic and structural insights. In this review, we discuss recent in silico structure-based and physics-based approaches currently employed to model protein–peptide interactions with a few examples of their applications.

1. INTRODUCTION

Protein–protein interactions (PPIs) (Panagiotis & Bonvin, 2013) are ubiquitous in biological systems as they are the building blocks for the complex molecular machinery that drives critical functions within the cell. Mutations and perturbations within the proteins that result in aberrant interactions often lead to a number of diseases and cancer (Stites, 1997). In many PPIs it has been shown that the interactions are dominated by short linear recognition motifs that are part of the larger protein (London, Movshovitz-Attias, & Schueler-Furman, 2010). It has been said that as much as 40% of interactions between proteins in eukaryotes are mediated by these short linear binding motifs (London, Raveh, & Schueler-Furman, 2013). Disrupting or modulating PPIs by means of standard drug design principles based on small-molecule inhibitors has proven difficult. Often protein–protein binding recognition sites are shallow and rather feature-less compared to "druggable" sites (Panagiotis & Bonvin, 2013). PPIs are often characterized by the cooperative formation of many weak interatomic interactions over an extended binding surface rather than by few strong interactions, as in the case of protein–drug binding. Biologically protein–protein recognition is also often highly regulated and environment dependent. For example, the formation of protein complexes is, in many cases, modulated by residue phosphorylation or substrate binding. The mechanism of regulated recognition involves conformational changes in the binding partners which are prevented or induced by external signals. This implies that protein–protein binding motifs can be labile and intrinsically flexible. Clearly, it may not be always possible to recapitulate the complexities of protein–protein recognition with small-molecule inhibitors (Araghi & Keating, 2016; Azzarito, Long, Murphy, & Wilson, 2013).

To address these challenges, there has been a growing interest in developing PPI inhibitors based on peptides or peptide derivatives designed so as to mimic the binding motif of one of the partners of the protein–protein complex one wishes to disrupt (Arkin, Tang, & Wells, 2014; Basse et al., 2013; Higueruelo et al., 2009; Labbé, Laconde, Kuenemann, Villoutreix, & Sperandio, 2013). A classical example is to attempt to inhibit a PPI using a peptide with the same sequence of the known or assumed protein–protein recognition motif. While sensible, this strategy often does not yield good inhibition. This can be due to a variety of factors. PPIs are often strengthened by secondary interactions not captured by individual peptides. Secondary interactions can also help induce the reorganization of the binding partners in their bound conformations. Furthermore, unlike their corresponding motifs as part of the protein scaffold, peptides in solution tend to assume a large variety of conformations with only few of them compatible with complexation. The entropic and intramolecular energy cost for the peptide to fold in the correct bound conformation adversely affects binding affinity. Furthermore, strong protein–peptide complexes often display interaction patterns different from the ones often encountered in protein–protein complexes. For example, protein–peptide interactions tend to be more hydrogen bond dependent than PPIs, usually involving the peptide backbone (London et al., 2010).

The study of protein–peptide interactions is also motivated by their biological significance. Peptides often serve as signaling vehicles as, for example, in the case of hormones and neurotransmitters (Fricker, 2012). Peptides in organisms often occur as part of the breakdown of proteins. One important example is the Amyloid beta (Abeta) peptide which results from the hydrolysis of the APP protein on the surface of neurons. Plaques formed by the aggregation of Abeta peptides are recognized as one of the hallmarks of Alzheimer's disease (Hardy & Allsop, 1991). Antimicrobial peptides (Pino-Angeles, Leveritt, & Lazaridis, 2016) are an important component of the innate immune system, and peptides presented and recognized on the surface of antigen-presenting cells by the fragmentation of foreign proteins are recognized as one of the main processes of the adaptive immune response.

The binding affinity between two molecules is measured quantitatively by the dissociation constant, K_d, or, equivalently, the binding constant, K_b (Gallicchio & Levy, 2011; Panagiotis & Bonvin, 2013). Binding constants for protein–peptide complexes can be obtained using a variety of experimental techniques (Day, Baird, Rich, & Myszka, 2002; Piehler, 2005; Shoemaker & Panchenko, 2007a; Tellinghuisen & Chodera, 2011; Tsiang et al., 2009). Models for protein–peptide binding are less well established. Bioinformatic approaches based, for example, on sequence homology and phylogenetic profiles are being used to search for likely interacting protein sequences (Shoemaker & Panchenko, 2007b). Structure-based approaches to infer PPI profiles on a genomic scale have also been proposed (Zhang et al., 2012). These tools are useful for qualitative predictions on a large scale. However, despite the encouraging progress in the area of host–guest and protein–ligand binding (Gallicchio et al., 2015, 2014; Kaus et al., 2015; Klimovich, Shirts, & Mobley, 2015; Lin & Roux, 2013; Mobley et al., 2007), the prediction of the structures of protein–peptide complexes and the *quantitative* prediction of their binding affinities are far from a solved problem. Efficiently predicting binding affinities of protein–peptide binding affinities from structural models would be an extremely important step in our overall understanding of the regulation of biological systems and critical for drug design and development. Structure-based peptide design based on PPIs is growing in commercial interest as computational methods as well as computer processing hardware have become more advanced (Audie & Swanson, 2013).

In this review, we present an overview of structure-based and physics-based computational approaches in current use by the research community to model protein–peptide interactions. The review does not attempt to be comprehensive; rather it is meant as a compendium of the main methodologies with examples of their applications.

2. THEORY AND METHODS

In this section we briefly review the physio-chemical theory of binding and some modeling technologies: binding free energy models and docking in current use.

2.1 Statistical Mechanics Formulation of Molecular Association

From a physio-chemical perspective the noncovalent binding of two molecular species, A and B, can be thought of as a reversible chemical process at equilibrium. The binding affinity is the probability of this interaction governed by the rates of association and disassociation, k_{on} and k_{off}, respectively:

$$A + B \underset{k_{\text{off}}}{\overset{k_{\text{on}}}{\rightleftarrows}} AB \tag{1}$$

At equilibrium, the rate of the forward reaction equals the rate of the reverse reaction, and therefore the equilibrium constant for binding K_b can be written as

$$K_b = \frac{[AB]/C^0}{([A]/C^0)([B]/C^0)} = \frac{1}{C^0}\frac{k_{on}}{k_{off}} \tag{2}$$

where C^0 is the standard state concentration which is often set to 1 M or 1 molecule/1668 Å^3. Eqs. (1) and (2) assume that all species involved can be treated as distinct chemical species. The standard binding free energy, ΔG_b^0, which is a thermodynamic measure of the binding affinity, is defined as

$$\Delta G_b^0 = -k_B T \ln K_b \tag{3}$$

where k_B is Boltzmann's constant. These equations, usually first introduced in undergraduate physical chemistry, are deceptive in their simplicity. The reality is that it is exceedingly difficult to compute the binding free energy between two proteins from first principles. The problem is inherently statistical in nature, having to do with the ensemble of conformations of the complex AB and their binding partners A and B when not bound. Statistical mechanics theory yields the following expression for the binding constant (Gallicchio & Levy, 2011; Gilson, Given, Bush, & McCammon, 1997)

$$K_b = \frac{C^0}{8\pi^2}\frac{Z_{N,AB} Z_N}{Z_{N,A} Z_{N,B}} \tag{4}$$

where Z_N is the configurational partition function of the solvent composed of N molecules and $Z_{N,AB}$, $Z_{N,A}$, and $Z_{N,B}$ are the configurational partition functions of the complex, protein A, and protein B, respectively. For example, the partition function for the molecule B

$$Z_{N,B} = \int e^{-\beta U(x_B, r_s)} dx_B dr_s \tag{5}$$

where the molecule is placed in some arbitrary position and orientation in solution and integration is carried over the $3n_B - 6$ internal degrees of freedom, x_B. n_B is the number of atoms in protein B, r_s represents the degrees of freedom of the solvent, and (Ux_B, r_s) is the potential energy of the solvent/solute system. The external degrees of freedom of molecule B correspond to three translational and three rotational coordinates (collectively denoted by ζ_B) that specify the position and orientation of molecule B with respect to the receptor A (Boresch, Tettinger, Leitgeb, & Karplus, 2003; Gallicchio &

Levy, 2011). The configurational partition function of the complex is then be written as

$$Z_{N,AB} = \int_{\text{bound}} e^{-\beta U(x_A, x_B, \zeta_B, r_s)} dx_A dx_B d\zeta_B dr_s \quad (6)$$

where the "bound" macrostate is any suitable structural definition of the complex based on the external coordinates ζ_B, such as the distance between the centers of mass of A and B (Gallicchio & Levy, 2012).

The statistical mechanics expression (4) forms the basis of many physics-based models of molecular binding in current use for the study of protein–drug binding equilibria (Gallicchio & Levy, 2011; Gilson et al., 1997; Guvench & MacKerell, 2009). Evaluation of (4) is daunting and complex, even for modern computers. Free energy methods (Chipot & Pohorille, 2007; Chodera et al., 2011; de Ruiter & Oostenbrink, 2011; Gallicchio, Lapelosa, & Levy, 2010) attempt to evaluate the ratio of partition functions in Eq. (4) either directly (Chang & Gilson, 2004; Kollman et al., 2000) or by free energy perturbation (FEP) approaches (Chodera et al., 2011; de Ruiter & Oostenbrink, 2011; Gallicchio et al., 2010) with varying degrees of physical rigor. Each approach comes with its pros and cons. We will discuss the most relevant of these methods as they apply to current applications in protein–peptide binding.

2.2 Pathway Free Energy Methods

Unlike end-point methods, such as MM/PB/GBSA and Mining Minima (Chang & Gilson, 2004; Kollman et al., 2000), discussed below, pathway free energy methods compute the difference in binding free energy between the unbound and bound states of the ligand–protein complex by linking them through a suitable thermodynamic path. This is equivalent to compute the ratio in Eq. (4) by averaging (Gallicchio & Levy, 2011; Gilson et al., 1997; Guvench & MacKerell, 2009). Pathway free energy methods are based on prescriptions to compute free energy differences along the thermodynamic path as ratios of partition functions, as such, do not require models of the absolute free energies of the end states. Often times the path is parameterized by a progress parameter λ ranging from 0 to 1. A λ-dependent potential energy function $U(x_A, x_B, \zeta_B, r_s;\lambda)$ is designed in such a way that at $\lambda = 0$ it represents the unbound state of the complex (when protein and ligand are not interacting), and at $\lambda = 1$ it represents the bound state of the complex. The reversible work along the path yields the binding free energy of the complex (de Ruiter & Oostenbrink, 2011).

Pathway free energy methods are broadly classified based on either physical or alchemical pathways (Deng & Roux, 2009). In physical pathway approaches the ligand is transferred from bulk solvent to the binding site following a physical path in space. In principle, the binding pathway can be parameterized along any of the six intermolecular degrees of freedom, ζ, representing the position and orientation of the ligand relative to the receptor. Although, in practice, to reduce conformational sampling volume, the ligand orientations are fixed and the pathway is described by a chosen distance measure parameterized by the progress parameter λ (Woo & Roux, 2005). The output of the calculation is the free energy along the pathway, referred to as the potential of mean force (PMF), $\Delta F(\zeta)$, defined as

$$\mathrm{e}^{-\beta\Delta F(\zeta)} = \frac{\int dx_{\mathrm{A}}dx_{\mathrm{B}}dr_{s}\mathrm{e}^{-\beta U(x_{\mathrm{A}},x_{\mathrm{B}},\zeta,r_{s})}}{\int dx_{\mathrm{A}}dx_{\mathrm{B}}dr_{s}\mathrm{e}^{-\beta U(x_{\mathrm{A}},x_{\mathrm{B}},\zeta^{*},r_{s})}}, \tag{7}$$

where ζ^{*} specifies an arbitrary position of the ligand in the solvent bulk sufficiently removed from the receptor so that it does not interact with it. The binding free energy, or equivalently the binding constant, is then estimated from the integral over the binding site region of the PMF:

$$K_{\mathrm{b}} = \mathrm{e}^{-\beta\Delta G_{b}^{0}} = \frac{C^{\circ}}{8\pi^{2}}\int_{\mathrm{site}} d\zeta \mathrm{e}^{-\beta\Delta F(\zeta)}, \tag{8}$$

where the integration is over the binding site volume as defined (Gallicchio & Levy, 2011; Gilson et al., 1997). The PMF for binding can be obtained by a variety of computational means including umbrella sampling (Gan & Roux, 2009) and metadynamics (Cavalli, Spitaleri, Saladino, & Gervasio, 2014). Nonequilibrium approaches can also be used to explore binding pathways and obtain binding affinity estimates (Nerattini, Chelli, & Procacci, 2016).

In alchemical approaches the ligand is not physically transferred to the binding site; rather the ligand is confined within the binding site region and the transformation between the bound and unbound states of the complex is carried out in so-called alchemical space by dialing in ligand–receptor interactions (Gallicchio & Levy, 2011). Computationally this is achieved by constructing a λ-dependent potential energy function whereby, for example, at $\lambda = 0$ the ligand is fully interacting with the receptor and at $\lambda = 1$ ligand–receptor interactions are turned off (alchemical decoupling). Due to the presence of the solvent, decoupling cannot normally be accomplished directly. Instead a thermodynamic cycle is employed whereby the binding free energy is obtained by the difference of the free energies of two separate decoupling calculations (double decoupling) (Boyce et al., 2009; Gilson et al., 1997). In one, the ligand–receptor and ligand–solvent interactions are turned off so that at the end state the ligand is effectively a ghost molecule freely moving in the binding site region, while the receptor interacts only with the solvent medium. This state is often referred to as the "vacuum" state of the ligand. The second transformation involves decoupling of the ligand from the solvent in the absence of the receptor to obtain the hydration free energy of the ligand (Chodera et al., 2011).

To our knowledge there has been only one report of the application of the double decoupling method to the calculation of absolute binding free energies of peptides (Gumbart, Roux, & Chipot, 2012). The decoupling transformations are too extensive and the corresponding free energies are too large to routinely obtain converged binding free energy estimates. An alchemical method based on implicit representations of the solvent exists which allows calculation of absolute free energies of binding using only one decoupling calculation (Gallicchio et al., 2010). Because they compute the binding free energy directly rather than as the difference of separate estimates whose errors add constructively, single-decoupling alchemical approaches could potentially be applied to peptide binding. Indeed single decoupling has been already applied to the binding of relatively large ligands (Gallicchio et al., 2014), and efforts to extend the method to peptide binding are under way (see below).

While challenging for absolute binding free energies, alchemical methods based on dual transformation legs are considered the state of the art for the accurate estimation of relative binding free energies (Wang et al., 2015). These methods are known in the field as FEP approaches, albeit somewhat confusingly, as in principle all pathway-based free energy methods are based on thermodynamic perturbation (Chipot & Pohorille, 2007; de Ruiter & Oostenbrink, 2011). The theoretical basis of relative binding free energy methods is similar to double decoupling except that the alchemical transformations involve mutating one ligand into another rather than completely decoupling them from the receptor/solvent environment. Methods of this kind are applicable to protein–peptide binding in cases when the transformations are sufficiently small, such as the mutation of an amino acid residue into another, and when they do not cause large conformational rearrangements (Rashid et al., 2013).

2.3 MM-PBSA, MM-GBSA

MM-PBSA (molecular mechanics/Poisson–Boltzmann surface area) and MM-GBSA (molecular mechanics/generalized Born surface area) are the two most common end-point methods applied to protein–ligand/protein–peptide interactions (de Ruiter & Oostenbrink, 2011; Genheden & Ryde, 2015). Another example of an end-point method is the Mining Minima method (Chang, Chen, & Gilson, 2007; Chen, Gilson, Webb, & Potter, 2010), which, because it has not been applied to protein–peptide binding, will not be further discussed. The advantages of end-point methods are that they only require the simulation of the protein–peptide complex and the free protein and free peptide states, thereby eliminating pathway steps. This is seemingly computationally efficient, but in practice accuracy is limited (de Ruiter & Oostenbrink, 2011). In more detail, the binding free energy ΔG_{bind} is estimated as

$$\Delta G_{\text{bind}} = G_{\text{AB}} - G_{\text{A}} - G_{\text{B}} \tag{9}$$

where the terms in the right-hand-side represent the free energy of the complex and the separate binding partners, respectively. In MM/(PBSA or GBSA) each free energy is estimated as (Genheden & Ryde, 2015)

$$G = \langle E_{\text{bond}} \rangle + \langle E_{\text{ele}} \rangle + \langle E_{\text{vdW}} \rangle + \langle G_{\text{pol}} \rangle + \langle G_{\text{np}} \rangle - TS \tag{10}$$

where E_{bond} is the energy of covalent interactions, E_{ele} is the energy of electrostatic interactions, E_{vdW} is the energy of van der Waals interactions, and G_{pol} and G_{np} are the polar and nonpolar contributions to the solvation free energy modeled by means of an implicit representation of the solvent (see below). T is the absolute temperature and S is the entropy estimated by normal-mode analysis. The solvent polarization energy G_{pol} term is calculated by solving the Poisson–Boltzmann equation either numerically (MM/PBSA) or using the generalized Born approximation (MM/GBSA), while the G_{np} is usually estimated from a linear relation to the solvent-accessible surface area. Energy terms are averaged over an ensemble of conformations sampled with either explicit solvation (more common) or implicit solvation. In principle, individual ensembles corresponding to each of the three states (AB, A, and B) should be obtained to evaluate the free energy of each state using Eq. (10). However, to simplify the calculation, in practice it is more common to use only the conformational ensemble of the complex to evaluate the free energies of all three species (Genheden & Ryde, 2015). In applications it is also common to neglect the entropic term (Brown & Muchmore, 2006).

2.4 Software for Protein–Peptide Docking

Increasingly peptides are becoming of major interest in the drug development field as it becomes more apparent that proteins interact through specific binding motifs that can be reduced to shortened peptides, and as discussed earlier, some proteins inherently react with small peptides. Protein substrates that are well known for their affinity to peptides include the MHC (major histocompatibility complex) used in immune recognition, SH3, and PDZ domains (Dagliyan, Proctor, D'Auria, Ding, & Dokholyan, 2011). The first step in understanding important protein–protein/peptide interactions is to obtain a structural model of the protein–peptide complex. Ideally structure determination can be accomplished by experimental means such as crystallography and NMR. However, this is not always possible, particularly for large libraries of peptides. Molecular docking can be a very useful tool in this respect (Meng, Zhang, Mezei, & Cui, 2011), especially when taking into account the wealth of structural data already available in databases such as the Protein Data Bank (Rose et al., 2011). There is a wide choice of software for protein–drug docking (Ewing, Makino, Skillman, & Kuntz, 2001; Friesner et al., 2004; Perryman, Santiago, Forli, Santos-Martins, & Olson, 2014; Verdonk, Cole, Hartshorn, Murray, & Taylor, 2003; Zhou, Felts, Friesner, & Levy, 2007) and rigid protein–protein docking (Gray et al., 2003; Kozakov, Brenke, Comeau, & Vajda, 2006; Pierce et al., 2014). Here we review docking software that is suitable for or has been used for protein–peptide docking. In a way, protein–peptide docking is more challenging than either protein–drug docking or rigid protein–protein docking. Protein–peptide docking involves more degrees of freedom than protein–drug docking (Tubert-Brohman, Sherman, Repasky, & Beuming, 2013). Adding to the complexity, the assumption of some level of rigidity common in protein–protein docking is simply not applicable to flexible peptides.

2.4.1 Local Docking: Peptide Docking to a Known Binding Site

In most instances, successful modeling of protein–peptide complexes depends on the prior structural knowledge of the protein–peptide complex or some homolog system. Local docking approaches exploit prior structural knowledge of a related system to produce structural predictions. In these situations, typically, many peptides are evaluated for a single receptor based on a structural template derived from a known structure. Once this initial template is established and peptides prepared, local docking software samples multiple conformations near the template and returns what it calculates are the possible structures. A few current local docking options are described below.

HADDOCK (High Ambiguity Driven protein–protein Docking) is a docking program based on experimental biochemical and biophysical interaction data (Dominguez, Boelens, & Bonvin, 2003). The information on interacting residues is provided by a group of restraints named Ambiguous Interaction Restraints (AIRs). These restraints are compiled from experimental information about the residues of the binding interface and their intermolecular interactions. For example, if using NMR titration data, residues involved in binding are distinguished from others by measuring the chemical shift perturbations upon complex formation. The AIRs are used to narrow down the possible conformations of the protein–peptide complex. Docked conformations are then scored based on electrostatic, van der Waals energy, and restraint energies.

FlexPepDock is a software that aims at modeling protein–peptide complexes that relies on two algorithms: (1) a refinement protocol (Raveh, London, & Schueler-Furman, 2010) and (2) and ab initio protocol (Raveh, London, Zimmerman, & Schueler-Furman, 2011). The refinement protocol, implemented within Rosetta, uses an initial coarse-grained model of a protein–peptide complex using PDB coordinates and approximate backbone coordinates for the peptide in the receptor binding site. The rigid peptide backbone is then optimized relative to the receptor using the Rosetta full atom energy function, the Rosetta coarse-grained energy function, Monte Carlo sampling with energy minimization, and a side-chain repacking protocol. The ab initio protocol, like the refinement protocol, requires the initial PDB coordinates of the receptor and the backbone coordinates of a bound peptide. The software then prepares a fragment library of trimer, pentamer, and nonamer backbone fragments that are extracted from the PDB based on sequence homology to the original peptide. Subsequently thousands of peptide models are generated from the original structure and the prepared fragment library. These models are then optimized using the refinement protocol described earlier. The resulting optimized structures are then clustered and ranked based on their energy scores.

GalaxyPepDock (Lee, Heo, Lee, & Seok, 2015) requires a protein structure in PDB format (900 amino acid maximum) and a FASTA peptide sequence (30 amino acid maximum) as input. Protein–peptide complex structure templates are then generated from the PepBind (Das, Sharma, Kumar, Krishna, & Mathur, 2013) database. Selected templates are then scored based on protein structure alignment and peptide sequence alignment using GalaxyTBM (Ko, Park, & Seok, 2012). Further optimization and refinement is completed using the GalaxyRefine (Heo, Park, & Seok, 2013) protocol which uses repetitive molecular mechanics relaxations to adjust backbone and side-chain conformations.

MedusaDock (Ding, Yin, & Dokholyan, 2010) is a flexible docking approach that sets out to solve the major problem of induced fit, where bound and unbound conformational states of the binding partners are different. MedusaDock simultaneously models peptide and receptor flexibility using sets of discrete rotamers. MedusaDock uses a novel algorithm that builds a rotamer library in situ during docking simulations. MedusaDock then uses the MedusaScore algorithm to guide the docking and rank the poses.

2.4.2 Global Docking: Peptide Docking to an Unknown Binding Site

Global Docking methods are the procedure of choice when the binding site is uncertain. Most often ensembles of peptide conformations are generated first, followed by rigid-body docking of peptide conformers to the surface of the target protein. The computational efficiency and accuracy of these methods are highly dependent on the size of the search area. Most protocols use a coarse-grained model of the receptor as a starting point to find viable binding sites and then move on to atomic resolution refinement to optimize selections. Two global docking protocols are discussed below.

The CABS-dock (Kurcinski, Jamroz, Blaszczyk, Kolinski, & Kmiecik, 2015) protocol is a modeling procedure based on the coarse-grained CABS protein model (Blaszczyk, Jamroz, Kmiecik, & Kolinski, 2013). The CABS-dock protocol begins with generating random peptide structures and randomly placing them on a surface of a sphere centered at the receptor's centroid. The peptide structures are then simulated using Replica Exchange Monte Carlo methods with replicas uniformly spread on a temperature scale. At the end of the replica search the replicas at the lowest temperatures are sampled for low-energy docked poses. Selected structures are then clustered together and selected models are then optimized using MODELLER (Eswar, Eramian, Webb, Shen, & Sali, 2008) with DOPE statistical potentials (Shen & Sali, 2006).

PepATTRACT (Schindler, de Vries, & Zacharias, 2015) is a flexible protein–peptide docking protocol that combines coarse-grained peptide docking to search for possible binding sites on the entire receptor surface along with atomistic flexible refinement. Peptide model structures are first generated from a given amino acid sequence followed by global rigid-body docking to the receptor surface using a coarse-grained force field implemented with ATTRACT (May & Zacharias, 2008). Possible solutions are ranked and those selected are first refined using iATTRACT (Schindler et al., 2015) and then optimized using molecular dynamics simulations with AMBER (Case et al., 2005).

3. SAMPLE APPLICATIONS

In this section we describe examples of works where the computational methodologies described above have been used in various combinations and for various purposes. This is not meant to be a complete and comprehensive overview, but rather a more insightful view of the computational approaches in use through application examples.

3.1 Predicting Protein–Peptide Interactions with Docking

Here we will present several recent papers that evaluate novel docking software for the purposes of predicting the location of likely binding sites and the conformation of the peptide in the complex.

In a 2015 paper Wabik, Kurcinski, and Kolinski (2015) used coarse-grained modeling with Monte Carlo temperature replica exchange, implemented with the CABS-dock web server, to evaluate how well the method was able to predict the complex formed between Troponin C and the end terminal helix of Troponin I whose interaction plays a role in the regulation of skeletal and cardiac muscle contraction. To measure success, the authors analyzed the root mean square deviation (RMSD) of the C_α atoms of predicted bound poses relative to available crystallographic structures. They also conducted protein–peptide contact analysis based on close proximity between residues. The method was able to place the peptide in the correct binding site within an overall RMSD of 2.5 Å compared to crystal structures. It was also shown that the software is also capable of locating alternative binding poses through cluster analysis of structural predictions.

Dokholyan, Buldyrev, Stanley, and Shakhnovich (1998) used all-atom temperature replica exchange discrete molecular dynamics (DMD) to predict the binding site and peptide conformations of multiple protein–peptide complexes including PDZ domain (1BFE), homer evh1 (1DDW), Src SH3 domain (1SRL), Keap1 (1X2J), and phospholipase A2 (1CL5), among others. The authors also set out to evaluate the effect conformational flexibility has on the accuracy of the model. Unlike traditional molecular dynamics (MD), which is based on continuous potentials, DMD implements ballistic equations of motion on discontinuous piece-wise constant pair potentials. DMD allows access to longer timescales than MD, especially for coarse-grained macromolecular representations. The DMD method was used to find the initial binding sites of the peptides and refinement was done utilizing MedusaDock. The effect of conformational flexibility was determined by varying different levels of receptor flexibility for each simulation such as (1) completely rigid, (2) flexible side chains with rigid backbone, or (3) flexible side chains and flexible backbone with an overall biasing potential for the native apo structure. In the initial simulations, peptides are allowed to randomly traverse the surface of the receptor. Hierarchical clustering is performed on peptide conformations that are within a certain cutoff distance of the receptor. Clusters with the lowest energy are scored with MedusaDock, and the receptor–peptide complex with the highest docking score is compared with the known crystal structure of the complex. For the 10 complexes studied, the RMSD values of the two most highly populated clusters from the known crystal structures ranged from 2.5 Å to about 37 Å. The lowest RMSD structures were from the data set that used flexible side chains. The authors also found that electrostatic interactions play a primary role in finding the correct binding site and orientation of the peptide. Simulations without electrostatic interactions yielded significantly poorer structural predictions. Overall the authors concluded that using a protocol that allows flexible side chains and includes electrostatic interactions greatly improves the protein–peptide complex prediction.

A 2013 study by Trellet, Melquiond, and Bonvin (2013) presents an optimized HADDOCK protocol for flexible protein–peptide docking that includes conformational selection and induced fit recognition. The authors tested their protocol on the petiDB dataset (London et al., 2010), which is a protein–peptide benchmark. Due to the fact that HADDOCK uses an experimental data-driven approach, the authors defined a binding surface on the receptor to drive the docking. HADDOCK was able to find acceptable solutions 72% of the time if backbone and side chains were made flexible, confirming previous studies (Dokholyan et al., 1998). In the case of a peptide that binds in an extended conformation, like in SH3 domains, HADDOCK was shown to have a 60% success rate. However, if the peptide binds in a secondary structure (helical, β-hairpin, and turns), HADDOCK is less successful at finding the correct binding pose and the failure rate is high. To amend this problem, the authors introduced restraints that held the peptides into the preferred, known bound conformation. This increased the success rate. HADDOCK's docking ability was also tested starting with solution conformations of the peptides. They applied the same restraining protocol and found that if the peptide and/or receptor does not change conformation significantly, the protocol has a good success rate. If the conformation does change upon binding, the protocol fails, which is to be expected.

3.2 Conformational and Energetic Analysis by Molecular Dynamics

There has been an increased interest in developing peptidomimetic small-molecule inhibitors especially in the area of histone-modifying and chromatin-binding proteins (Tortorici et al., 2013). Many systems involved in histone modification have implications in cancer, viral infections, genetic diseases, etc. However, it has been shown that PPIs are difficult to mimic using small molecules (Azzarito et al., 2013). One of the many obstacles that small molecules face are their inability to bind ambiguous binding pockets or make all the critical interactions that a larger molecule would be otherwise be able to (Araghi & Keating, 2016). Another route to synthesizing PPI inhibitors is to engineer antibodies. However, antibodies are large molecules that are limited to extracellular targets (Araghi & Keating, 2016). The solution is to develop small, optimized peptides that can bind to the target receptor binding site. The solution is never as easy as it seems. Most PPIs are mediated through secondary structures. As evidenced by Trellet et al. (2013) when developing their HADDOCK protocol, many peptides did not yield good docked poses unless restrained to a bound form usually consisting of an α-helical secondary structure. Therefore one of the first steps in designing small peptide inhibitors is to test the affinity of multiple peptide sequences, of varying length and amino acid content, that mimic the binding residues of the original protein site. Computational studies can assist experimental binding assays by first identifying sequences that retain or form the correct secondary structure needed for binding as well as evaluating amino acid mutations and other structural changes that might result in improved binding. This is often done with MD simulations of the protein–peptide complex as well as simulations of the free peptide in solution.

As an example, in 2013, Tortorici et al. (2013) investigated a series of small peptides derived from the "SNAG" domain of SNAIL1 that inhibits lysine-specific demethylase (LSD1) which, with the corepressor CoREST, selectively demethylates Lys4me/me2 on histone H3. It was shown through the analysis of seven crystal structures of SNAIL-derived peptides bound to LSD1/CoREST that the N-terminal domain residues of the peptide were bound in a helical structure. However, CD spectra performed on the peptides in solution showed that they exist as mostly random conformers. The authors turned to computational MD to corroborate their experimental data. They performed molecular dynamics simulations in explicit solvent with AMBER (Case et al., 2005) as well as temperature replica exchange MD simulations to explore the peptide conformations in solution. True to the experimental data, MD simulations revealed that the peptides exist as random structures, showing only transient secondary structures with an α-helical propensity of only about 7%. The authors used this evidence to suggest that the peptides do indeed exist as disordered conformers in solution and only form secondary structure upon binding.

As most peptides are disordered in solution but usually need to be folded into some kind of secondary structure to bind, one new direction of interest is to engineer conformationally restrained peptides with the expectation that this will increase binding affinity. One active area of research concerns the notion of "stapled" peptides. These restrained peptides are expected to have a more favorable binding energy than that of nonrestrained peptides due to reduced entropic cost of adapting to the bound, often helical, structure (Miles et al., 2016). These constraints can be manufactured in different ways: using hydrocarbon linkages, disulfide bridges, incorporating amino acids with helical propensity, and others (Miles et al., 2016). Several recent papers have studied the effects of staples on protein–peptide binding (Gerona-Navarro et al., 2011; Lama et al., 2013; Miles et al., 2016). One computational study is discussed below.

Eukaryotic mRNAs are capped at their 5′-ends by a m^7GTP structure that is involved with regulation of translation, RNA stability, RNA nuclear transport, and splicing. eIF4E is a translation initiating factor that is frequently overexpressed in different cancers causing unregulated mRNA translation (Banko et al., 2005). Lama et al. (2013) set out to design novel stapled peptide inhibitors of eIF4E that are derived from eIF4G; eIF4G is an another transcription initiation factor that is required by eIF4E to recruit the 40S ribosomal subunit to the cap. Their initial analysis consisted of MD simulations to sample the protein–peptide bound conformations initiated from X-ray crystal structures and replica exchange molecular dynamics of the unbound peptides to sample possible conformations. The dominant structures from MD simulations of the complex were mostly α-helical and consistent with the crystal structure. On the other hand, REMD of the free peptides showed dominant structures that were mostly disordered. To improve helicity in solution, the authors introduced i, i + 4 and i, i + 3 hydrocarbon staples at various attachment locations. They found that the staple increased, as expected, the peptide helicity in solution to up to 50%. However, this did not necessarily improve binding affinity. Several of the staples led to decreased flexibility of residues critical for binding. In some cases the staples also increased the population of conformational states of the peptides incompatible with binding. Changing the placement of the staple to improve critical residue flexibility as well as to remove spurious interactions was shown to increase binding affinity, although often by mechanisms not apparent from the crystal structure. The authors concluded that staples do increase helicity of free peptides in solution; however, this often leads to peptides with decreased flexibility that may not be able to overcome the reorganizational energy required to bind in conformations similar to those of the bound form crystal structure. However, careful placement of staples and considerations of critical residues may improve or modify native interactions resulting in increased binding affinity.

3.3 Modeling Key Interactions with Mutagenic Analysis to Improve Affinity and Selectivity of Binding

Another avenue to pursue in protein–peptide interactions, besides selecting an amino acid sequence and restraining that sequence into a native/bound secondary structure, is to focus on the critical interactions themselves, perhaps building a linear peptide that makes the necessary interactions without needing to fold into a complex secondary structure needed to bind. In addition, there are many receptors such as GPCRs (G-protein-coupled receptors) that bind to many different ligands and proteins differing in only a small number of key interactions. The concept of selectivity is extremely important for drug design and drug efficacy. One way to accomplish this is to model protein–peptide interactions and then measure the effect of mutations on binding affinity.

Methods that go beyond docking have been implemented to study binding selectivity. As an illustration we describe here work on the Kv1.3 potassium channel, which is an important target for the treatment of autoimmune diseases such as rheumatoid arthritis and multiple sclerosis

(Rashid et al., 2013). Selectivity is a key in this system as possible peptidomimetic drugs can bind with high affinity to other unwanted Kv channels, in particular the Kv1.1 channel (Rashid et al., 2013). The ShK peptide has spawned several potent analogs with improved selectivity. However, these analogs contain nonnatural amino acids and/or adducts. With this in mind, Rashid et al. (2013) set out to identify a selective ShK analog composed of only natural amino acids by means of FEP and thermodynamic integration (TI) calculations to evaluate the binding free energy changes associated with introducing mutations. Rashid et al. also evaluated binding affinities to the Kv1.1 and Kv1.3 receptors using PMF methods (Deng & Roux, 2009).

Previously published experimental alanine mutagenesis data (Pennington et al., 1996; Rauer, Pennington, Cahalan, & Chandy, 1999) provided a starting point for the computational work of Rashid et al. Binding mode analysis revealed that Lys18 makes a critical electrostatic interactions with a Glu residue of the Kv1.1 receptor, while no such interaction exists with for the Kv1.3 receptor. The authors then proceeded to use FEP and TI methods to evaluate the free energy change of the K18A mutation as well as the PMF method to calculate the absolute binding free energy of the mutant peptide to each of the Kv channels as well as wild type. Rashid et al. (2013) were able to obtain binding free energies that agreed well with experimental data. Whole-cell patch-clamp assays revealed that a mutant they selected displayed a 124-fold increase in selectivity for the Kv1.3 channel over the Kv1.1. The authors also tested for off-target toxic effects on cardiac potassium channels and found no such issue. Here the authors exemplified the use of computational methods in the area of drug lead optimization to discover critical interactions that aid in the understanding of binding mechanism that can be used to enhance selectivity.

A further example of a computational molecular dynamics study elucidating molecular recognition mechanism and assisting in discovering possible novel analogs is a study published in 2014 that evaluated human parathyroid hormone through mutagenesis and interaction analysis (Jiang, Jiande, Lei, Chen, & Jin, 2014). Human parathyroid hormone is an 84-amino acid peptide that regulates extracellular calcium homeostasis and bone remodeling by binding to the extracellular domain (ECD) of the PTH1R G-protein-coupled receptor leading to activation. X-ray crystallographic data show that hPTH interacts with this receptor with its N-terminal (1–34) residues, with residues 1–14 interacting with regions of the PTH1R transmembrane domain and residues 15–34 bind with high affinity and specificity to the ECD (Jiang et al., 2014). In this study, the crystal structure of PTH1R/hPTH (15–34) was extracted from the Protein Data Bank (3C4M) and was subsequently prepared with the Discovery Studio software by Accelrys Inc. The free energy of mutation was calculated with Discovery Studio using CHARMM and a generalized Born implicit solvent model. Explicit solvent MD simulations on mutant-receptor binding were subjected to trajectory analysis to evaluate structural properties including RMSD, root mean square fluctuations, and radius of gyration. Experimental site-directed mutagenesis showed that Leu24 and Leu28 are critical residues that, when substituted with Glu, cause a binding affinity decrease of about 4000- and 1600-fold, respectively. The calculated free energy of mutants was predicted to be quite large in conjunction with an unstable conformation in agreement with experimental data. One the other hand, a triple mutant, R25Q-K26E-K27L, was predicted to have a stable helical secondary structure and enhanced binding free energy. The authors then assessed the bioactivity of this triple mutant by investigating the expression levels of RANKL and OPG genes using RT-PCR. Wild-type PTH stimulates the transcription of the RANKL gene and inhibits OPG. The triple mutant increased the RANKL/OPG ratio by 30 compared to wild type. Hence, as a result of the computational studies, the authors identified a triple mutant that was experimentally shown to have high bioactivity compared to wild type. This study as well as others discussed above shows the crucial role that computational chemistry can play in elucidating molecular recognition mechanisms and enhancing drug discovery.

3.4 Distinguishing Binders from Nonbinders

The high-throughput screening of ligand databases is an established procedure to find new drug leads (Lu et al., 2015). There is a strong interest to do the same with libraries of peptides. As peptides are becoming of increasing interest in the field of drug discovery, researchers are taking up the challenge and developing new methodologies and protocols to deal with these new, more complex systems. As with ligands, a good starting point is to test novel protocols in terms of their ability to distinguish known binders from nonbinders from a library of peptides. Due to their atomistic resolution and computational efficiency, MM-PBSA and MM-GBSA tend to be the methods of choice in this area. Here we present a couple of examples of how these methods have been used to computationally screen peptide libraries.

In a 2006 paper Hou, Chen, McLaughlin, Lu, and Wang (2006) used MM-PBSA to determine relative binding free energies of peptides taken from a peptide library containing potential binders to human Abl SH3 domain. A total of 35 peptides were tested, with 15 binders and 20 nonbinders. MM-PBSA was able to distinguish the binders from the nonbinders with relative binding affinities with a r^2 correlation value of 0.82 relative to experimental values. This correlation was obtained with the inclusion of the entropy term $-T\Delta S$, which without the correlation coefficient is lowered to 0.72. The electrostatic term was shown to be the interaction energy component providing the strongest correlation with the experiments. In this system electrostatics also provided the largest contribution to the binding affinity. These findings suggest that potentially these methods would be less successful for systems in which electrostatics is less important, such as when the binding interface is mostly hydrophobic.

In the study of Mena-Ulecia, Vergara-Jaque, Poblete, Tiznado, and Caballero (2014) MM-GBSA was used to study the binding affinities of a series of short peptides against the catalytic site of protein kinase A. The standard approach was used with molecular dynamics using the NAMD software package (Phillips et al., 2005) and parameters from the CHARMM27 force field (Best et al., 2012; MacKerell et al., 1998) and MM-GBSA using the NAMD 2.8 (Phillips et al., 2005). To reduce computational cost the entropy term was not computed. This approach was justified by the observation that other studies on similar systems had concluded that the entropy term did not contribute significantly to the binding affinity and/or would not vary appreciably across the peptide library. The authors reported high correlation between experimental free energy values and computational values (r^2= 0.988). The authors also concluded that the most influential term in the binding free energy is the electrostatic term, again showing that the success of this model is intimately tied to the electrostatic interactions present.

In this paper the impact of specific mutations of a small peptide (LRRASLG) named Kemptide, after the designer in the original study (Kemp, 1990), was also investigated. All of the mutations focused on the two critical arginine residues in the sequence, either with mutations to a similarly charged residue (K or H) or an alanine. They were able to conclude that one arginine residue was more important than the other based on binding affinity (one made more favorable hydrogen bonds) and this agreed with the affinity of the R to A mutants which were, not surprisingly, less favorable but in a way that agreed that one arginine was more critical to binding than the other. The authors concluded that MM-GBSA is a simple, computationally inexpensive method that can be used to investigate free energies as well as analyze atomistic interactions that influence the binding affinity. This should be taken with a grain of salt. As we noted, this is true for interactions that are highly dependent on electrostatic interactions and may not be the case for more subtle interactions.

3.5 Protein–Peptide Binding Affinities

The ultimate goal of computational modeling of macromolecular binding is the estimation of binding free energies from physical principles. This is a very complex computational problem. After several decades of efforts, useful progress is being obtained in the area of protein–drug binding (Gallicchio et al., 2014; Wang et al., 2015). There are few examples of applications of binding free energy methods to protein–peptide bindings. They are reviewed here.

The binding of a series of four-residue phosphorylated peptides to an SH2 regulatory domain has been studied by the Roux group using a free energy PMF methodology (Gan & Roux, 2009; Woo & Roux, 2005). The procedure involves several perturbation steps, each contributing to the overall free energy of binding. The general idea at the basis of this and later works by this group (Gumbart et al., 2012; Gumbart, Roux, & Chipot, 2013; Jo, Jiang, & Roux, 2015) is to reduce the amount of conformational sampling required by first restraining the peptide in solution, then carrying out the PMF calculation with the restrained peptide, and then finally releasing the restraints with the peptide bound to the receptor. The first step is the localization of the peptide in a fixed position in solution in front and at some distance from the receptor site to ensure an unobstructed binding path. In the second step, the conformation of the peptide is restrained so as to mimic its known or assumed conformation in the bound state. The PMF for binding has then been calculated along the peptide–receptor site distance with umbrella sampling using a series of harmonic restraining potentials distributed along the binding path.

Src homology 2 (SH2) domains are regulatory domains present in many enzymes. They recognize specific motifs containing a phosphorylated tyrosine. The free energy calculations by Gan and Roux (2009) recapitulated known trends in peptide recognition by five SH2 domains (Lck, Grb2, Cbl, p85αN, and Stat1) yielding binding free energies in good agreement with experimental measurements. The calculations elucidated the key electrostatic interactions responsible for the recognition of the phosphate group, and the more subtle interactions responsible for the modulation of binding by flanking residues. The free energy decomposition afforded by the method revealed the important role played by the conformational reorganization free energy as measured by the difference of the free energies for imposing and releasing the conformational restraints. The authors acknowledged that the success of the calculations were in part due to the rigidity of the receptor. Indeed, later applications of the PMF method to more flexible receptors encountered conformational sampling challenges (Gumbart et al., 2013).

Gumbart et al. (2012) compared the relative merits of PMF and alchemical routes to the calculation of the binding free energy between a proline-rich peptide and the SH3 domain of tyrosine kinase Abl. The double-decoupling alchemical approach requires the separate evaluations of the decoupling free energies of the peptide in solution and in the receptor site. Due to the size of ligand, these can be quite large and difficult to converge. On the other hand, the PMF approach models the direct transfer of the peptide from bulk water to the protein receptor site and is not as challenging. In either case the prior restraining of the peptide to limit the amount of conformational sampling necessary helped in reaching convergence of the calculations. As they should, the alchemical and PMF approaches yielded a similar estimate of the protein–ligand binding free energy.

With implicit solvation the distinction between the PMF and alchemical approaches becomes more blurred. The BEDAM method (Gallicchio et al., 2010), for example, is based on the effective binding energy function $u(x)$ which can be interpreted as either the adiabatic free energy change for turning on ligand–receptor interactions or the adiabatic free energy change for transferring the ligand from the solvent phase to the receptor. An early application of the BEDAM single-decoupling approach has been to study protein–peptide interactions in an mRNA silencing complex involved in fragile X syndrome (Di Marino, D'Annessa, Tancredi, Bagni, & Gallicchio, 2015), one of the primary causes of autism in children (Bagni & Oostra, 2013). As discussed previously, control of mRNA translation is a crucial step in many cellular processes as dysregulation leads to cancer and other diseases (Lama et al., 2013). In addition to the eIF4E–eIF4G binding discussed above, eIF4E is known to share the same binding site with several other proteins and peptides called 4E-BPs that are implicated in many processes including the development of synaptic plasticity (Banko et al., 2005). CYFIP1 (cytoplasmic FMRP interacting protein 1) is a member of the 4E-BP family that, together with FMRP (fragile X mental retardation protein), binds to eIF4E acting as an inhibition complex (Napoli et al.,

2008). Mutations in FMRP inhibit complex formation with CYFIP1 and eIF4E, leading to fragile X syndrome (Bagni & Oostra, 2013; Rubeis et al., 2014). Di Marino et al. (2015) set out to computationally model the key amino acid sequence of CYFIP1 that binds to eIF4E with the hope that the information gained from this study will lead to the engineering of a peptide inhibitor that can serve to reregulate eIF4E if FMRP is defective.

As no crystal structures of the eIF4E–CYFIP1 complex exist, the authors built homology models of the complex of eIF4E with a CYFIP1-derived peptide based on the crystal structures of 4E-BP- and eIF4G-derived peptides bound to eIF4E. Interestingly, molecular dynamics simulations with explicit solvation started from these homology models, resulting in unbinding and unfolding of the CYFIP1-derived peptide. It was therefore concluded that CYFIP1 must bind eIF4E in a unique binding mode that differed from the other 4E-BPs. The authors set out to predict the structure of eIF4E bound to a CYFIP1-derived peptide starting from a docking model of the eIF4E–CYFIP1 complex obtained using the HADDOCK program (Dominguez et al., 2003; Marino et al., 2015). In this model the CYFIP1 peptide was bound to a different site than that shared by 4E-BPs. To ensure exhaustive sampling during the conformational search, the authors used multidimensional replica exchange molecular dynamics with the BEDAM protocol (Gallicchio et al., 2010), normally used to predict binding free energies (Gallicchio et al., 2014). BEDAM simulations showed a distinct binding mode for CYFIP1 that was near the canonical binding site but altered by significant rotation. MD simulations on this new binding mode showed stability in the peptide secondary structure and stable binding to eIF4E in contrast to earlier MD where the complex lacked stability. The authors were also able to ascertain unique binding interactions with the canonical binding site that are not present in the other peptides. Overall the work of Di Marino et al. (2015) shows that high-level modeling using different techniques can help gain invaluable insights in protein–peptide complexation. The work also validated the BEDAM binding free energy method as a conformational search tool in difficult cases such as this where peptide folding is coupled to binding.

The convergence properties of single-decoupling protocols such as BEDAM, which are already well established for protein–drug binding (Gallicchio, 2012; Gallicchio et al., 2014; Lapelosa, Gallicchio, & Levy, 2012), are currently being studied in the context of protein–peptide binding. The benchmark system being evaluated (Fig. 1) consists of a series of cyclic peptide inhibitors of the interaction of HIV integrase with the lens

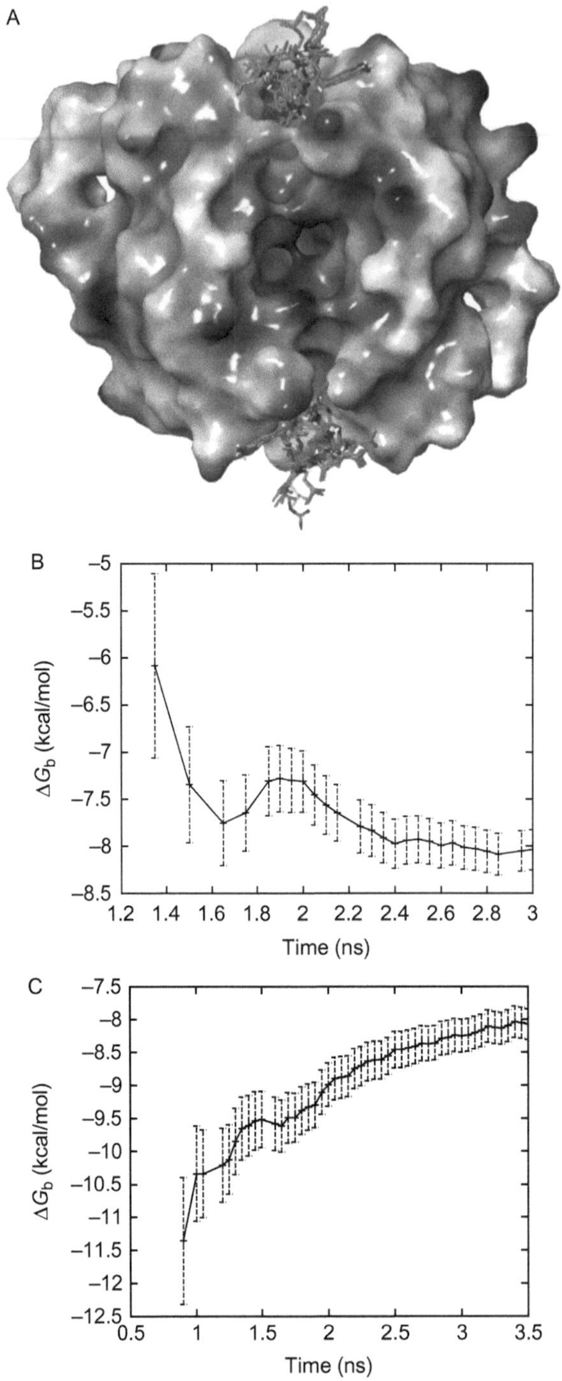

Fig. 1 (A) Representation of the complex catalytic core domain of HIV integrase (*blue* (*dark gray* in the print version), *red* (*gray* in the print version), and *gray*) with two cyclic peptides (*green* (*light gray* in the print version)). BEDAM convergence plots (computed binding free energy as a function of simulation time) for two cyclic peptides: SLKIDNLD (B, wild-type sequence) and the ALKIDNMD double mutant (C).

epithelium-derived growth factor host protein, which has been shown to be critical for the integration of the viral genome into the host chromosome (Rhodes et al., 2011; Tsiang et al., 2009). As shown in Fig. 1 the BEDAM calculation for the wild-type sequence converges to a value in reasonable agreement with the experimental measurements (Rhodes et al., 2011); however, a double-mutant sequence fails to converge within a similar simulation time. The simulations were started from the crystal structure of the complex with the wild-type peptide (Rhodes et al., 2011), so it is likely that the difference in convergence times is related to the fact that the double-mutant peptide is undergoing slow conformational reorganization to assume a binding pose different from the one assumed by the wild-type peptide.

4. CONCLUDING REMARKS

The full treatment of protein–peptide equilibria from first principles remains for the most part out of reach from methods and software. The size and flexibility of peptides and the complexity of the interactions they form with protein receptors present seemingly insurmountable obstacles. Despite this, very useful computational tools have been devised to model various aspects of peptide design from different point of views; molecular docking is being used to predict the likely structures of protein–peptide complexes. Molecular dynamics and energetic scoring methods provide a quantitative representation of the strength of interactions and the entropic costs associated with peptide folding. Advanced binding free energy and conformational sampling protocols are improving, also thanks to ever faster computers. It is our hope that this necessarily incomplete and imperfect overview will be useful to researchers to navigate the range of approaches available to study and understand protein–peptide complexation.

REFERENCES

Araghi, R. R., & Keating, A. E. (2016). Designing helical peptide inhibitors of protein–protein interactions. *Current Opinion in Structural Biology*, *39*, 27–38.

Arkin, M. R., Tang, Y., & Wells, J. A. (2014). Small-molecule inhibitors of protein-protein interactions: Progressing toward the reality. *Chemistry & Biology*, *21*(9), 1102–1114.

Audie, J., & Swanson, J. (2013). Advances in the prediction of protein–peptide binding affinities: Implications for peptide-based drug discovery. *Chemical Biology & Drug Design*, *81*(1), 50–60.

Azzarito, V., Long, K., Murphy, N. S., & Wilson, A. J. (2013). Inhibition of [alpha]-helix-mediated protein-protein interactions using designed molecules. *Nature Chemistry*, *5*(3), 161–173.

Bagni, C., & Oostra, B. A. (2013). Fragile X syndrome: From protein function to therapy. *American Journal of Medical Genetics. Part A*, *161*(11), 2809–2821.

Banko, J. L., Poulin, F., Hou, L., DeMaria, C. T., Sonenberg, N., & Klann, E. (2005). The translation repressor 4E-BP2 is critical for eIF4F complex formation, synaptic plasticity, and memory in the hippocampus. *The Journal of Neuroscience, 25*(42), 9581–9590.

Basse, M. J., Betzi, S., Bourgeas, R., Bouzidi, S., Chetrit, B., Hamon, V., … Roche, P. (2013). 2P2Idb: A structural database dedicated to orthosteric modulation of protein–protein interactions. *Nucleic Acids Research, 41*(D1), D824–D827.

Best, R. B., Zhu, X., Shim, J., Lopes, P. E. M..Mittal, J., & Feig, M.MacKerell, A. D., Jr.. (2012). Optimization of the additive CHARMM all-atom protein force field targeting improved sampling of the backbone ϕ, ψ and side-chain $\chi 1$ and $\chi 2$ dihedral angles. *Journal of Chemical Theory and Computation, 8*(9), 3257–3273.

Blaszczyk, M., Jamroz, M., Kmiecik, S., & Kolinski, A. (2013). CABS-fold: Server for the de novo and consensus-based prediction of protein structure. *Nucleic Acids Research, 41*(W1), W406–W411.

Boresch, S., Tettinger, F., Leitgeb, M., & Karplus, M. (2003). Absolute binding free energies: A quantitative approach for their calculation. *The Journal of Physical Chemistry B, 107*, 9535–9551.

Boyce, S. E., Mobley, D. L., Rocklin, G. J., Graves, A. P., Dill, K. A., & Shoichet, B. K. (2009). Predicting ligand binding affinity with alchemical free energy methods in a polar model binding site. *Journal of Molecular Biology, 394*, 747–763.

Brown, S. P., & Muchmore, S. W. (2006). High-throughput calculation of protein-ligand binding affinities: Modification and adaptation of the MM-PBSA protocol to enterprise grid computing. *Journal of Chemical Information and Modeling, 46*, 999–1005.

Case, D. A., Cheatham, T. E. III.Darden, T., Gohlke, H., Luo, R., Merz, K., Jr, … Woods, R. J. (2005). The Amber biomolecular simulation programs. *Journal of Computational Chemistry, 26*, 1668–1688.

Cavalli, A., Spitaleri, A., Saladino, G., & Gervasio, F. L. (2014). Investigating drug–target association and dissociation mechanisms using metadynamics-based algorithms. *Accounts of Chemical Research, 48*(2), 277–285.

Chang, C. A., Chen, W., & Gilson, M. K. (2007). Ligand configurational entropy and protein binding. *Proceedings of the National Academy of Sciences of the United States of America, 104*, 1534–1539.

Chang, C. E., & Gilson, M. K. (2004). Free energy, entropy, and induced fit in host-guest recognition: Calculations with the second-generation mining minima algorithm. *Journal of the American Chemical Society, 126*, 13156–13164.

Chen, W., Gilson, M. K., Webb, S. P., & Potter, M. J. (2010). Modeling protein-ligand binding by mining minima. *Journal of Chemical Theory and Computation, 6*(11), 3540–3557.

Chipot, C., & Pohorille, A. (2007). *Free energy calculations. Theory and applications in chemistry and biology. Springer series in chemical physics*. Berlin, Heidelberg: Springer.

Chodera, J. D., Mobley, D. L., Shirts, M. R., Dixon, R. W., Branson, K., & Pande, V. S. (2011). Alchemical free energy methods for drug discovery: Progress and challenges. *Current Opinion in Structural Biology, 21*, 150–160.

Dagliyan, O., Proctor, E. A., D'Auria, K. M., Ding, F., & Dokholyan, N. V. (2011). Structural and dynamic determinants of protein-peptide recognition. *Structure, 19*(12), 1837–1845.

Das, A. A., Sharma, O. P., Kumar, M. S., Krishna, R., & Mathur, P. P. (2013). PepBind: A comprehensive database and computational tool for analysis of protein–peptide interactions. *Genomics, Proteomics & Bioinformatics, 11*(4), 241–246.

Day, Y. S., Baird, C. L., Rich, R. L., & Myszka, D. G. (2002). Direct comparison of binding equilibrium, thermodynamic, and rate constants determined by surface-and solution-based biophysical methods. *Protein Science, 11*(5), 1017–1025.

Deng, Y., & Roux, B. (2009). Computations of standard binding free energies with molecular dynamics simulations. *The Journal of Physical Chemistry. B, 113*, 2234–2246.

de Ruiter, A., & Oostenbrink, C. (2011). Free energy calculations of protein–ligand interactions. *Current Opinion in Chemical Biology*, *15*(4), 547–552.

Di Marino, D., D'Annessa, I., Tancredi, H., Bagni, C., & Gallicchio, E. (2015). A unique binding mode of the eukaryotic translation initiation factor 4E for guiding the design of novel peptide inhibitors. *Protein Science*, *24*, 1370–1382.

Ding, F., Yin, S., & Dokholyan, N. V. (2010). Rapid flexible docking using a stochastic rotamer library of ligands. *Journal of Chemical Information and Modeling*, *50*(9), 1623–1632.

Dokholyan, N. V., Buldyrev, S. V., Stanley, H. E., & Shakhnovich, E. I. (1998). Discrete molecular dynamics studies of the folding of a protein-like model. *Folding and Design*, *3*(6), 577–587.

Dominguez, C., Boelens, R., & Bonvin, A. M. (2003). HADDOCK: A protein-protein docking approach based on biochemical or biophysical information. *Journal of the American Chemical Society*, *125*(7), 1731–1737.

Eswar, N., Eramian, D., Webb, B., Shen, M.-Y., & Sali, A. (2008). Protein structure modeling with MODELLER. In B. Kobe, M. Guss, & T. Huber (Eds.), *Structural proteomics: high-throughput methods:* (Vol. 426, pp. 145–159). Humana Press.

Ewing, T. J. A..Makino, S., Skillman, A. G., & Kuntz, I. D. (2001). Dock 4.0: Search strategies for automated molecular docking of flexible molecule databases. *Journal of Computer-Aided Molecular Design*, *15*(5), 411–428.

Fricker, L. D. (2012). Neuropeptides and other bioactive peptides: From discovery to function. In *Colloquium series on neuropeptides:* (Vol. 1, pp. 1–122). San Rafael, CA: Morgan & Claypool Life Sciences.

Friesner, R. A., Banks, J. L., Murphy, R. B., Halgren, T. A., Klicic, J. J., Mainz, D. T., … Shenkin, P. S. (2004). Glide: A new approach for rapid, accurate docking and scoring. 1. Method and assessment of docking accuracy. *Journal of Medicinal Chemistry*, *47*(7), 1739–1749.

Gallicchio, E. (2012). Role of ligand reorganization and conformational restraints on the binding free energies of DAPY non-nucleoside inhibitors to HIV reverse transcriptase. *Computational Molecular Bioscience*, *2*, 7–22.

Gallicchio, E., Chen, H., Chen, H., Fitzgerald, M., Gao, Y., He, P., … Levy, R. M. (2015). BEDAM binding free energy predictions for the SAMPL4 octa-acid host challenge. *Journal of Computer-Aided Molecular Design*, *4*, 315–325.

Gallicchio, E., Deng, N., He, P., Perryman, A. L., Santiago, D. N., Forli, S., … Levy, R. M. (2014). Virtual screening of integrase inhibitors by large scale binding free energy calculations: The SAMPL4 challenge. *Journal of Computer-Aided Molecular Design*, *28*, 475–490.

Gallicchio, E., Lapelosa, M., & Levy, R. M. (2010). Binding energy distribution analysis method (BEDAM) for estimation of protein-ligand binding affinities. *Journal of Chemical Theory and Computation*, *6*, 2961–2977.

Gallicchio, E., & Levy, R. M. (2011). Recent theoretical and computational advances for modeling protein-ligand binding affinities. *Advances in Protein Chemistry and Structural Biology*, *85*, 27–80.

Gallicchio, E., & Levy, R. M. (2012). Prediction of SAMPL3 host-guest affinities with the binding energy distribution analysis method (BEDAM). *Journal of Computer-Aided Molecular Design*, *25*, 505–516.

Gan, W., & Roux, B. (2009). Binding specificity of SH2 domains: Insight from free energy simulations. *Proteins: Structure, Function, and Bioinformatics*, *74*(4), 996–1007.

Genheden, S., & Ryde, U. (2015). The MM/PBSA and MM/GBSA methods to estimate ligand-binding affinities. *Expert Opinion on Drug Discovery*, *10*(5), 449–461.

Gerona-Navarro, G., Mujtaba, S., Frasca, A., Patel, J., Zeng, L., Plotnikov, A. N., … Zhou, M.-M. (2011). Rational design of cyclic peptide modulators of the transcriptional coactivator CBP: A new class of p53 inhibitors. *Journal of the American Chemical Society*, *133*(7), 2040–2043.

Gilson, M. K., Given, J. A., Bush, B. L., & McCammon, J. A. (1997). The statistical-thermodynamic basis for computation of binding affinities: A critical review. *Biophysical Journal, 72*, 1047–1069.

Gray, J. J., Moughon, S., Wang, C., Schueler-Furman, O., Kuhlman, B., Rohl, C. A., & Baker, D. (2003). Protein–protein docking with simultaneous optimization of rigid-body displacement and side-chain conformations. *Journal of Molecular Biology, 331*(1), 281–299.

Gumbart, J. C., Roux, B., & Chipot, C. (2012). Standard binding free energies from computer simulations: What is the best strategy? *Journal of Chemical Theory and Computation, 9*(1), 794–802.

Gumbart, J. C., Roux, B., & Chipot, C. (2013). Efficient determination of protein–protein standard binding free energies from first principles. *Journal of Chemical Theory and Computation, 9*(8), 3789–3798.

Guvench, O., & MacKerell, A. D. (2009). Computational evaluation of protein-small molecule binding. *Current Opinion in Structural Biology, 19*, 56–61.

Hardy, J., & Allsop, D. (1991). Amyloid deposition as the central event in the aetiology of Alzheimer's disease. *Trends in Pharmacological Sciences, 12*, 383–388.

Heo, L., Park, H., & Seok, C. (2013). GalaxyRefine: Protein structure refinement driven by side-chain repacking. *Nucleic Acids Research, 41*(Web Server Issue), W384–W388.

Higueruelo, A. P., Schreyer, A., Bickerton, G. R. J., Pitt, W. R., Groom, C. R., & Blundell, T. L. (2009). Atomic interactions and profile of small molecules disrupting protein–protein interfaces: The TIMBAL database. *Chemical Biology & Drug Design, 74*(5), 457–467.

Hou, T., Chen, K., McLaughlin, W. A., Lu, B., & Wang, W. (2006). Computational analysis and prediction of the binding motif and protein interacting partners of the Abl SH3 domain. *PLoS Computational Biology, 2*(1), e1.

Jiang, Y., Jiande, G., Lei, J., Chen, Y., & Jin, J. (2014). Molecular dynamics and bioactivity of a novel mutated human parathyroid hormone. *Tropical Journal of Pharmaceutical Research, 13*(4), 511–518.

Jo, S., Jiang, W., & Roux, B. (2015). Quantifying protein-protein binding energy and entropy using molecular dynamics simulations. *Biophysical Journal, 108*(2), 41a.

Kaus, J. W., Harder, E., Lin, T., Abel, R., McCammon, J. A., & Wang, L. (2015). How to deal with multiple binding poses in alchemical relative protein-ligand binding free energy calculations. *Journal of Chemical Theory and Computation, 11*(6), 2670–2679.

Kemp, B. E. (1990). *Peptides and protein phosphorylation*. Boca Raton, FL: CRC Press.

Klimovich, P. V., Shirts, M. R., & Mobley, D. L. (2015). Guidelines for the analysis of free energy calculations. *Journal of Computer-Aided Molecular Design, 29*(5), 397–411.

Ko, J., Park, H., & Seok, C. (2012). GalaxyTBM: Template-based modeling by building a reliable core and refining unreliable local regions. *BMC Bioinformatics, 13*(1), 1–8.

Kollman, P. A., Massova, I., Reyes, C., Kuhn, B., Huo, S., Chong, L., … Cheatham, T. E. (2000). Calculating structures and free energies of complex molecules: Combining molecular mechanics and continuum models. *Accounts of Chemical Research, 33*, 889–897.

Kozakov, D., Brenke, R., Comeau, S. R., & Vajda, S. (2006). PIPER: An FFT-based protein docking program with pairwise potentials. *Proteins: Structure, Function, and Bioinformatics, 65*(2), 392–406.

Kurcinski, M., Jamroz, M., Blaszczyk, M., Kolinski, A., & Kmiecik, S. (2015). CABS-dock web server for the flexible docking of peptides to proteins without prior knowledge of the binding site. *Nucleic Acids Research, 43*(W1), W419–W424.

Labbé, C. M., Laconde, G., Kuenemann, M. A., Villoutreix, B. O., & Sperandio, O. (2013). iPPI-DB: A manually curated and interactive database of small non-peptide inhibitors of protein–protein interactions. *Drug Discovery Today, 18*(19), 958–968.

Lama, D., Quah, S. T., Verma, C. S., Lakshminarayanan, R., Beuerman, R. W., Lane, D. P., & Brown, C. J. (2013). Rational optimization of conformational effects induced by hydrocarbon staples in peptides and their binding interfaces. *Scientific Reports, 3*, 3451.

Lapelosa, M., Gallicchio, E., & Levy, R. M. (2012). Conformational transitions and convergence of absolute binding free energy calculations. *Journal of Chemical Theory and Computation, 8*, 47–60.

Lee, H., Heo, L., Lee, M. S., & Seok, C. (2015). GalaxyPepDock: A protein–peptide docking tool based on interaction similarity and energy optimization. *Nucleic Acids Research, 43*(W1), W431–W435.

Lin, Y.-L., & Roux, B. (2013). Computational analysis of the binding specificity of Gleevec to Abl, c-Kit, Lck, and c-Src tyrosine kinases. *Journal of the American Chemical Society, 135*(39), 14741–14753.

London, N., Movshovitz-Attias, D., & Schueler-Furman, O. (2010). The structural basis of peptide-protein binding strategies. *Structure, 18*(2), 188–199.

London, N., Raveh, B., & Schueler-Furman, O. (2013). Peptide docking and structure-based characterization of peptide binding: From knowledge to know-how. *Current Opinion in Structural Biology, 23*(6), 894–902.

Lu, G., Lomonosova, E., Cheng, X., Moran, E. A., Meyers, M. J., Le Grice, S. F., … Tavis, J. E. (2015). Hydroxylated tropolones inhibit hepatitis B virus replication by blocking viral Ribonuclease H activity. *Antimicrobial Agents and Chemotherapy, 59*(2), 1070–1079.

MacKerell, A. D., Jr..Bashford, D., Bellott, M., Dunbrack, R. L., Jr..Evanseck, J. D., Field, M. J., … Karplus, M. (1998). All-atom empirical potential for molecular modeling and dynamics studies of proteins. *The Journal of Physical Chemistry B, 102*(18), 3586–3616.

Marino, D. D., Rubeis, S. D., Achsel, T., Chillemi, G., Tramontano, A., & Bagni, C. (2015). MD and docking studies reveal that the functional switch of CYFIP1 is mediated by a butterfly-like motion. *Journal of Chemical Theory and Computation, 11* (7), 3401–3410.

May, A., & Zacharias, M. (2008). Energy minimization in low-frequency normal modes to efficiently allow for global flexibility during systematic protein–protein docking. *Proteins: Structure, Function, and Bioinformatics, 70*(3), 794–809.

Mena-Ulecia, K., Vergara-Jaque, A., Poblete, H., Tiznado, W., & Caballero, J. (2014). Study of the affinity between the protein kinase PKA and peptide substrates derived from kemptide using molecular dynamics simulations and MM/GBSA. *PLoS One, 9*(10), e109639.

Meng, X.-Y., Zhang, H.-X., Mezei, M., & Cui, M. (2011). Molecular docking: A powerful approach for structure-based drug discovery. *Current Computer-Aided Drug Design*, 7(2), 146.

Miles, J. A., Yeo, D. J., Rowell, P., Rodriguez-Marin, S., Pask, C. M., Warriner, S. L., … Wilson, A. J. (2016). Hydrocarbon constrained peptides—Understanding preorganisation and binding affinity. *Chemical Science*, 7(6), 3694–3702.

Mobley, D. L., Graves, A. P., Chodera, J. D., McReynolds, A. C., Shoichet, B. K., & Dill, K. A. (2007). Predicting absolute ligand binding free energies to a simple model site. *Journal of Molecular Biology, 371*, 1118–1134.

Napoli, I., Mercaldo, V., Boyl, P. P., Eleuteri, B., Zalfa, F., Rubeis, S. D., … Bagni, C. (2008). The fragile X syndrome protein represses activity-dependent translation through CYFIP1, a new 4E-BP. *Cell, 134*(6), 1042–1054.

Nerattini, F., Chelli, R., & Procacci, P. (2016). II. Dissociation free energies in drug–receptor systems via nonequilibrium alchemical simulations: Application to the FK506-related immunophilin ligands. *Physical Chemistry Chemical Physics, 18*, 15005–15018.

Panagiotis, L. K., & Bonvin, A. M. (2013). On the binding affinity of macromolecular interactions: Daring to ask why proteins interact. *Journal of the Royal Society Interface, 10*(79), 20120835.

Pennington, M. W., Mahnir, V. M., Khaytin, I., Zaydenberg, I., Byrnes, M. E., & Kem, W. R. (1996). An essential binding surface for ShK toxin interaction with rat brain potassium channels. *Biochemistry, 35*(51), 16407–16411.

Perryman, A. L., Santiago, D. N., Forli, S., Santos-Martins, D., & Olson, A. J. (2014). Virtual screening with AutoDock Vina and the common pharmacophore engine of a low diversity library of fragments and hits against the three allosteric sites of HIV integrase: Participation in the SAMPL4 protein–ligand binding challenge. *Journal of Computer-Aided Molecular Design, 28*, 1–13.

Phillips, J. C., Braun, R., Wang, W., Gumbart, J., Tajkhorshid, E., Villa, E., … Schulten, K. (2005). Scalable molecular dynamics with NAMD. *Journal of Computational Chemistry, 26*(16), 1781–1802.

Piehler, J. (2005). New methodologies for measuring protein interactions in vivo and in vitro. *Current Opinion in Structural Biology, 15*(1), 4–14.

Pierce, B. G., Wiehe, K., Hwang, H., Kim, B.-H., Vreven, T., & Weng, Z. (2014). ZDOCK server: Interactive docking prediction of protein–protein complexes and symmetric multimers. *Bioinformatics, 30*(12), 1771–1773.

Pino-Angeles, A., Leveritt, J. M., III, & Lazaridis, T. (2016). Pore structure and synergy in antimicrobial peptides of the magainin family. *PLoS Computational Biology, 12*(1).

Rashid, M. H., Heinzelmann, G., Huq, R., Tajhya, R. B., Chang, S. C., Chhabra, S., … Kuyucak, S. (2013). A potent and selective peptide blocker of the Kv1. 3 channel: Prediction from free-energy simulations and experimental confirmation. *PLoS One, 8*(11), e78712.

Rauer, H., Pennington, M., Cahalan, M., & Chandy, K. G. (1999). Structural conservation of the pores of calcium-activated and voltage-gated potassium channels determined by a sea anemone toxin. *Journal of Biological Chemistry, 274*(31), 21885–21892.

Raveh, B., London, N., & Schueler-Furman, O. (2010). Sub-angstrom modeling of complexes between flexible peptides and globular proteins. *Proteins: Structure, Function, and Bioinformatics, 78*(9), 2029–2040.

Raveh, B., London, N., Zimmerman, L., & Schueler-Furman, O. (2011). Rosetta FlexPepDock ab-initio: Simultaneous folding, docking and refinement of peptides onto their receptors. *PLoS One, 6*(4), e18934.

Rhodes, D. I., Peat, T. S., Vandegraaff, N., Jeevarajah, D., Newman, J., Martyn, J., … Deadman, J. J. (2011). Crystal structures of novel allosteric peptide inhibitors of HIV integrase identify new interactions at the LEDGF binding site. *ChemBioChem, 12*(15), 2311–2315.

Rose, P. W., Beran, B., Bi, C., Bluhm, W. F., Dimitropoulos, D., Goodsell, D. S., … Bourne, P. E. (2011). The RCSB protein data bank: Redesigned web site and web services. *Nucleic Acids Research, 39*(Suppl. 1), D392–D401.

Rubeis, S. D., He, X., Goldberg, A. P., Poultney, C. S., Samocha, K., Cicek, A. E., … Buxbaum, J. D. (2014). Synaptic, transcriptional and chromatin genes disrupted in autism. *Nature, 515*(7526), 209–215.

Schindler, C. E. M..de Vries, S. J..Zacharias, M. (2015). Fully blind peptide-protein docking with pepATTRACT. *Structure, 23*(8), 1507–1515.

Shen, M.-Y., & Sali, A. (2006). Statistical potential for assessment and prediction of protein structures. *Protein Science, 15*(11), 2507–2524.

Shoemaker, B. A., & Panchenko, A. R. (2007). Deciphering protein–protein interactions. Part I. Experimental techniques and databases. *PLoS Computational Biology, 3*(3), e42.

Shoemaker, B. A., & Panchenko, A. R. (2007b). Deciphering protein–protein interactions. Part II. Computational methods to predict protein and domain interaction partners. *PLoS Computational Biology, 3*(4), e43.

Stites, W. E. (1997). Protein-protein interactions: Interface structure, binding thermodynamics, and mutational analysis. *Chemical Reviews, 97*(5), 1233–1250.

Tellinghuisen, J., & Chodera, J. D. (2011). Systematic errors in isothermal titration calorimetry: Concentrations and baselines. *Analytical Biochemistry, 414*(2), 297–299.

Tortorici, M., Borrello, M. T., Tardugno, M., Chiarelli, L. R., Pilotto, S., Ciossani, G., ... Mattevi, A. (2013). Protein recognition by short peptide reversible inhibitors of the chromatin-modifying LSD1/CoREST lysine demethylase. *ACS Chemical Biology, 8*(8), 1677–1682.

Trellet, M., Melquiond, A. S., & Bonvin, A. M. (2013). A unified conformational selection and induced fit approach to protein-peptide docking. *PLoS One, 8*(3), e58769.

Tsiang, M., Jones, G. S., Hung, M., Mukund, S., Han, B., Liu, X., ... Geleziunas, R. (2009). Affinities between the binding partners of the HIV-1 integrase dimer-lens epithelium-derived growth factor (IN dimer-LEDGF) complex. *Journal of Biological Chemistry, 284*(48), 33580–33599.

Tubert-Brohman, I., Sherman, W., Repasky, M., & Beuming, T. (2013). Improved docking of polypeptides with glide. *Journal of Chemical Information and Modeling, 53*(7), 1689–1699. Retrieved from 23800267. (PMID:).

Verdonk, M. L., Cole, J. C., Hartshorn, M. J., Murray, C. W., & Taylor, R. D. (2003). Improved protein–ligand docking using gold. *Proteins: Structure, Function, and Bioinformatics, 52*(4), 609–623.

Wabik, J., Kurcinski, M., & Kolinski, A. (2015). Coarse-grained modeling of peptide docking associated with large conformation transitions of the binding protein: Troponin I fragment–Troponin C system. *Molecules, 20*(6), 10763–10780.

Wang, L., Wu, Y., Deng, Y., Kim, B., Pierce, L., Krilov, G., ... Abel, R. (2015). Accurate and reliable prediction of relative ligand binding potency in prospective drug discovery by way of a modern free-energy calculation protocol and force field. *Journal of the American Chemical Society, 137*, 2695–2703.

Woo, H.-J., & Roux, B. (2005). Calculation of absolute protein-ligand binding free energy from computer simulations. *Proceedings of the National Academy of Sciences of the United States of America, 102*, 6825–6830.

Zhang, Q. C., Petrey, D., Deng, L., Qiang, L., Shi, Y., Thu, C. A., ... Honig, B. (2012). Structure-based prediction of protein-protein interactions on a genome-wide scale. *Nature, 490*(7421), 556–560.

Zhou, Z., Felts, A. K., Friesner, R. A., & Levy, R. M. (2007). Comparative performance of several flexible docking programs and scoring functions: Enrichment studies for a diverse set of pharmaceutically relevant targets. *Journal of Chemical Information and Modeling, 47*, 1599–1608.

CHAPTER THREE

Structural Plasticity in Globins: Role of Protein Dynamics in Defining Ligand Migration Pathways

C. Estarellas*, L. Capece†, C. Seira*, A. Bidon-Chanal*, D.A. Estrin†, F.J. Luque*,[1]

*Institute of Biomedicine (IBUB), Faculty of Pharmacy and Food Science, University of Barcelona, Santa Coloma de Gramenet, Spain
†Universidad de Buenos Aires, Facultad de Ciencias Exactas y Naturales, INQUIMAE-CONICET, Ciudad Universitaria, Ciudad de Buenos Aires, Argentina
[1]Corresponding author: e-mail address: fjluque@ub.edu

Contents

1. Introduction 60
   1.1 Structure–Dynamics–Function Relationships in Globins 62
2. Ligand Migration Pathways in the Globin Superfamily: Structure–Function Relationships 63
   2.1 Myoglobin 63
   2.2 Hemoglobin 65
   2.3 *Cerebratulus lacteus* Mini-Hb 66
   2.4 Truncated Hbs 67
   2.5 Protoglobin 72
3. The Coordination of the Heme Group and Its Role in Ligand Binding 73
   3.1 Neuroglobin 73
4. Conclusions 76
Acknowledgments 77
References 77

Abstract

Globins are a family of proteins characterized by the presence of the heme prosthetic group and involved in variety of biological functions in the cell. Due to their biological relevance and widespread distribution in all kingdoms of life, intense research efforts have been devoted to disclosing the relationships between structural features, protein dynamics, and function. Particular attention has been paid to the impact of differences in amino acid sequence on the topological features of docking sites and cavities and to the influence of conformational flexibility in facilitating the migration of small ligands through these cavities. Often, tunnels are carved in the interior of globins, and ligand

Advances in Protein Chemistry and Structural Biology, Volume 105
ISSN 1876-1623
http://dx.doi.org/10.1016/bs.apcsb.2016.07.002

exchange is regulated by gating residues. Understanding the subtle intricacies that relate the differences in sequence with the structural and dynamical features of globins with the ultimate aim of rationalizing the thermodynamics and kinetics of ligand binding continues to be a major challenge in the field. Due to the evolution of computational techniques, significant advances into our understanding of these questions have been made. In this review we focus our attention on the analysis of the ligand migration pathways as well as the function of the structural cavities and tunnels in a series of representative globins, emphasizing the synergy between experimental and theoretical approaches to gain a comprehensive knowledge into the molecular mechanisms of this diverse family of proteins.

1. INTRODUCTION

Globins are a wide family of proteins that share a specific globular fold able to accommodate in its interior the heme prosthetic group. This family includes archetypical proteins such as myoglobin (Mb) and hemoglobin (Hb), which have been used as suitable models to illustrate the detailed connections between structure, energy landscape, protein dynamics, and the functional role of the different protein substates. The importance of Mb and Hb in protein science cannot be underestimated, as it has been recognized by the fact that Mb can be considered to be "the hydrogen atom of biology" (Frauenfelder, McMahon, & Fenimore, 2003). Nevertheless, the number of novel globins discovered in the last years has been growing, as reflected in the incorporations of new globins discovered during the last decade, such as truncated Hbs, protoglobin (Pgb), neuroglobin (Ngb), and cytoglobin (Ctg), which have served to disclose the intrinsic complexity underlying the apparent simplicity of the common globin fold (Tejero & Gladwin, 2014; Vinogradov & Moens, 2008; Wajcman, Kiger, & Marden, 2009).

Globins pertain to the class of alpha-domain proteins, where the protein skeleton is formed by a bundle of eight helical segments, usually labeled A–H. The length of the helices varies considerably, as noted in the shorter and larger helices in Mb, which involves 7 and 28 residues, respectively (Fig. 1). The helical segments adopt a 3-over-3 arrangement through the assistance of connecting loops, giving rise to a globular protein with a highly hydrophobic core that contains the heme-binding cavity. The iron in the heme group is coordinated to the four equatorial porphyrin nitrogens and to the conserved proximal histidine F8 (HisF8) in one of the axial positions, leading to the characteristic pentacoordinated (5c) state. Although the globin

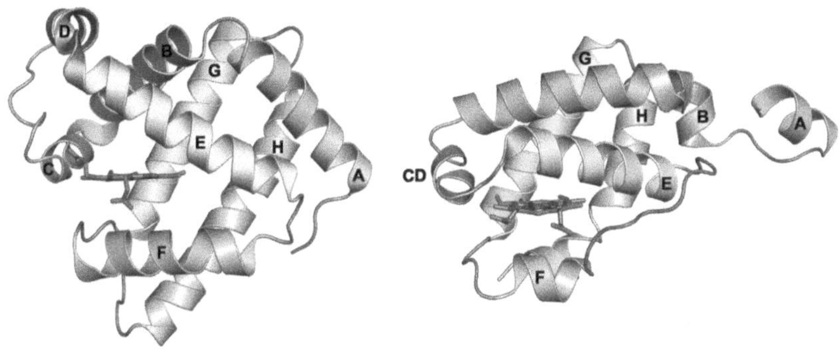

Fig. 1 *Left*: Representation of the typical 3-over-3 helical fold present in globins illustrated by Mb (PDB ID: 2W6X). *Right*: Representation of the 2-over-2 helical fold found in truncated HbN (PDB ID: 1IDR).

fold is highly evolutionarily conserved, it is worth noting that the sequence homology among these proteins can be as low as 16%, thus providing a molecular basis to the diversity of biological roles played in different organisms (Laberge & Yonetani, 2007).

Most of the functions associated with globins are related to the binding of external ligands to the heme group. Typically, exogenous ligands are small gaseous molecules, such as O_2, CO, and NO, that bind to the sixth coordination site (ie, the axial distal position). The affinity for a given ligand is determined by the equilibrium constant of the binding process, which is related to the kinetic rate constants of ligand association (k_{on}) and ligand dissociation (k_{off}). Globins have evolved to implement different strategies to modulate ligand affinity. The association process is mainly determined by the presence of docking sites and/or tunnels, where the transition of the ligand between cavities or the passage through tunnels, often regulated by gating residues, defines the barriers for ligand migration to the heme cavity. On the other hand, the dissociation process is primarily determined by the strength of protein–ligand interactions, which implies an energetic barrier much higher than the migration from the heme cavity to the bulk solvent. In turn, the strength of the protein–ligand interaction may be regulated in several ways, involving both distal and proximal effects. Distal effects are generally related to the formation of hydrogen-bond interactions between the ligand and distal residues, such as HisE7 in Mb or TyrB10 in the truncated HbN. Noteworthy, this is not a static factor, as the intrinsic plasticity of the distal cavity may permit the adoption of different conformational states and/or distinct hydrogen-bond patterns, which give rise to significant changes in the ligand-binding affinity. Other strategies are related to the proximal His, since the relative position of HisF8 to the heme plane and hydrogen bonding with other residues may also affect the ligand affinity.

As mentioned earlier, ligands bind to the vacant sixth coordination position of the heme. However, in several globins, the distal site is occupied by a protein residue belonging to the E helix, typically the HisE7 residue, forming the hexacoordinated (6c) species. Nevertheless, other residues may also coordinate to the iron, depending on the globin. For example, in the case of *Chlamydomonas eugametos* trHb and *Synechocystis* sp. truncated Hb, TyrB10 and HisE10 are the residues that coordinate to the heme axial position. This adds an additional level to regulate binding affinity, because in hexacoordinated globins the bound residue needs to be cleaved in order to form the pentacoordinated state, and hence the modulation of the 6c $\leftrightarrows$ 5c equilibrium results in an indirect mechanism of ligand-binding regulation.

1.1 Structure–Dynamics–Function Relationships in Globins

A challenging question is to disclose the relationships between the structural features arising from differences in the sequence content and the biochemical properties of globins. This knowledge is relevant for understanding the physiological role in the cell, since globins may encompass a wide variety of roles, such as O_2 transport and storage, enzymatic activities, oxygen sensing, and nitric oxide dioxygenase (NOD) activity. How does the structure of globins be evolutionarily designed to allow the migration of ligands? What specific residues play a crucial role in determining the thermodynamics and kinetics of ligand binding? How is the migration of ligands through cavities and tunnels assisted by protein dynamics? The answer to these questions requires the understanding of the nature of the dynamical motions that affect not only the side chains of the residues that shape cavities and tunnels (ie, gating residues), but also the flexibility of the protein backbone, and the possible competition between small ligands and water molecules present in the distal site, which could influence the ligand affinity.

These fundamental questions have been the subject of intense experimental and theoretical efforts since the discovery of the 3D structure of Mb around 60 years ago (Kendrew et al., 1958). Since then, Mb has been extensively studied as the prototypical member of the globin family. The internal cavities of Mb were considered as "packing defects," thus underestimating the importance of the cavities related to the function of proteins. However, the development of experimental techniques, such as time-resolved X-ray crystallography (Cho et al., 2010; Schmidt et al., 2005;

Schotte et al., 2003; Tomita et al., 2009), the crystallization under high Xe pressure (Pesce, Milani, Nardini, & Bolognesi, 2008; Sanctis et al., 2004; Tilton, Kuntz, & Petsko, 1984), and the structural and dynamical analysis afforded by extended molecular simulations (Abbruzzetti et al., 2013; Arroyo-Máñez et al., 2011; Capece et al., 2013), has shed light into the role of protein plasticity in assisting ligand migration.

The aim of this review is try to illustrate the role of structural fluctuations in assisting the migration of small ligands through the protein interior of globins. Due to the abundant literature, this goal is attained by examining a reduced set of representative cases, which nevertheless illustrate the variety of molecular factors implicated in ligand binding that provide a basis to gain insight into the subtle intricacies of structure–dynamics–function relationships in this family of proteins. Accordingly, in the following, emphasis will be made on the relation between the topology of inner cavities, the conformational flexibility of proteins, and the energetics of ligand migration.

2. LIGAND MIGRATION PATHWAYS IN THE GLOBIN SUPERFAMILY: STRUCTURE–FUNCTION RELATIONSHIPS

2.1 Myoglobin

Several research groups, using different theoretical techniques, have studied the ligand migration in Mb, showing how the ligands move across the Xe ligand-docking sites with small barriers between them (Banushkina & Meuwly, 2005; Bossa et al., 2005; Bossa, Anselmi, & Di Nola, 2004; Cohen, Arkhipov, Braun, & Schulten, 2006; Nutt & Meuwly, 2004; Ruscio et al., 2008), as revealed by time-resolved crystallography (Cho et al., 2010; Tomita et al., 2009).

In the distal site HisE7 is one of the most relevant structural features, not only for the energetic stabilization afforded to the heme-bound O_2, but also because of the ability to carry out a conformational flip, which orients this residue toward the bulk solvent, thus defining the HisE7 gate (Fig. 2). We have examined the oxygen migration through enhanced sampling techniques considering the HisE7 residue in the open and closed conformations (Boechi et al., 2013). We observed the expected shift from the closed to the open conformation upon protonation of HisE7, but more importantly, noted a significant difference between the conformations of the two neutral histidine tautomers. Further, a low energy barrier to ligand migration was found considering both open and closed conformations of HisE7.

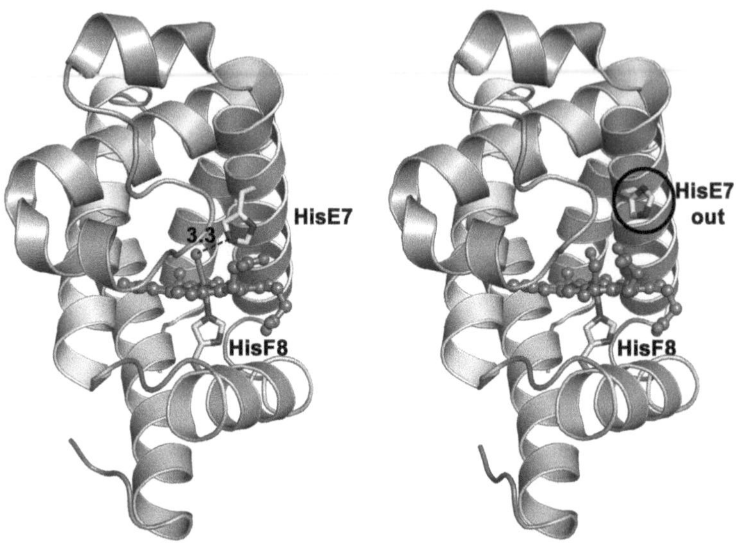

Fig. 2 Schematic representation of HisE7 gate hypothesis. *Left*: X-ray structure of Mb depicts HisE7 at 3.3 Å of the oxy-heme group, showing the closed conformation of this gate (PDB ID: 1VXF). *Right*: X-ray structure of Mb under extreme conditions (pH 4, $T=4°C$) displays the open conformation of the HisE7 gate (PDB ID: 1SPE).

In particular, the results show that even in the closed conformation, the HisE7 gate does not create a large barrier to oxygen migration and permits oxygen entry with only a small rotation of the imidazole side chain and movement of the E helix.

An interesting finding is that in the open conformation Mb may accommodate an additional oxygen molecule, which would agree with the major effective concentration of Mb. Additional studies were performed for the mutated variants HisE7Ala and HisE7Trp, which represent the fastest and slowest association rate constants, respectively (Olson, Soman, & Phillips, 2007). The mutation to Trp blocks completely the migration of O_2 to the distal side, but the mutation to Ala shows a deeper well compared to the wild-type Mb. This well is created due to a hydrophobic pocket in the tunnel. From the analysis of these results, we conclude that in this case the ligand migration was guided by a hydrophobic effect rather than by the classical open/close gating.

Finally, let us note that experimental studies point out that the preferential ligand migration pathway involves the distal E7 gate (Salter et al., 2012), which is estimated to be the dominant (>75%) migration pathway in Mb. Nevertheless, other works have identified a series of residues that can regulate the HisE7 gate, such as His64, Phe43, and Phe46, since mutation to small residues shows an increase in the rate association constant of O_2 (Scott, Gibson, & Olson, 2001).

2.2 Hemoglobin

The human Hb is a complex member of the globin family, formed by a tetrameric system composed of two α and two β chains, which acts as O_2 storage and carrier in humans. Ligand migration in Hb is not characterized as well as in Mb, since the presence of the two different chains in the quaternary assembly affords the structural basis for subtle regulation on the mechanism of ligand migration (Birukou, Schweers, & Olson, 2010; Birukou, Soman, & Olson, 2011; Mouawad, Marechal, & Perahia, 2005). A recent theoretical study using carbon monoxide as a probe (Lucas & Guallar, 2012) proposed that the ligand can find until four paths (Fig. 3). This work examined the R- and T-states of Hb considering both α and β chains independently. For the R-state in the α subunit, all the four paths are explored by at least 17% of the total (40) simulations. However, the ligand never explored a pathway involving the proximal site, as noted in the case of Mb. This may be explained by the fact that the residue Phe138 in Mb is

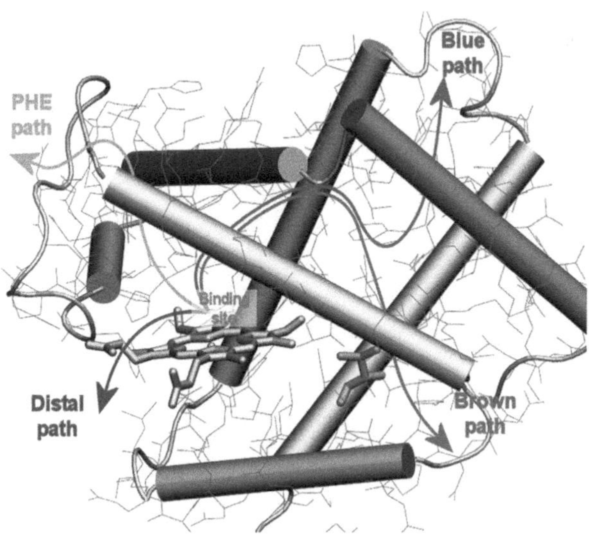

Fig. 3 Schematic representation of the carbon monoxide exits passages in human Hb. *Reprinted from Lucas, F., & Guallar, V. (2012). An atomistic view on human hemoglobin carbon monoxide migration processes.* Biophysical Journal, 102, *887–896.*

replaced by Val132 in Hb. In contrast, for the β chain in the R-state, 95% of the simulations revealed that the ligand migration path takes place through the distal site. With regard to the T-state, the behavior observed for α and β chains resembles the results obtained for the R-state, that is, the α chain shows a larger variability in the migration pathways, while in the β chain the most populated path corresponds to the distal site at least in 60% of the simulations. Additionally, when comparison is made between the migration paths in R- and T-states, it can be observed that the distal path is largely populated in the R-state. This fact could be related to the location of the E helix, which is more displaced in the R-state, thus facilitating the migration path through the distal side.

2.3 *Cerebratulus lacteus* Mini-Hb

The neuronal mini-Hb of *Cerebratulus lacteus* (CerHb) is the smallest functional globin known with 109 residues. The shorter sequence affects the typical 3-over-3 structural fold, since the N-terminal helix A is not present, the C-terminal H helix is shorter, but in contrast there is a larger GH region (Bolognesi, Bordo, Rizzi, Tarricone, & Ascenzi, 1997; Geuens et al., 2004). The function of CerHb is related to O_2 storage and releases to the axons and brain tissue during periods of anoxic conditions (Vandergon, Riggs, Gorr, Colacino, & Riggs, 1998).

In its distal site CerHb contains the TyrB10–GlnE7 pair. Globins that display these residues show very high O_2 affinity, and especially very low dissociation rates, as reported for *Ascaris suum* Hb and *Mycobacterium tuberculosis* truncated HbN, which exhibit O_2 dissociation constants (k_{off}) of 0.004 and 0.02 s^{-1}, respectively (Pesce et al., 2001; Yeh, Couture, Ouellet, Guertin, & Rousseau, 2000). This can be attributed to the strong hydrogen bonds formed by these residues with the heme-bound O_2. Interestingly, in CerHb, the O_2 affinity is moderate, and the O_2 dissociation constant is larger than in the previous cases (200–600 s^{-1}) (Pesce et al., 2004). At large extent, this behavior arises from the presence of Thr48 at position E11, since mutagenesis studies indicated that the ThrE11→Val mutation triggers a huge increase in the ligand affinity, with an eightfold increase in k_{on} and 1000-fold decrease in k_{off}. In this context, molecular simulations disclosed the existence of two conformations in the distal cavity with differential O_2 affinity. In the high-affinity conformation, TyrB10 is hydrogen bonded to O_2, whereas in the low-affinity conformation TyrB10 interacts with ThrE11, and the ligand stabilization is lost (Martí et al., 2006).

Another crucial structural feature in CerHb is the long apolar tunnel (~10 Å) that connects the heme distal site with the solvent thought the C-terminal ends. This tunnel was characterized experimentally obtaining the crystal structure of CerHb, which showed the presence of two Xe sites (Pesce et al., 2011). This tunnel is related to the high O_2 association rate constant (230 $\mu M^{-1}\ s^{-1}$) in comparison with other invertebrate globins that contains the Tyr–Gln active pair, which normally displays association rate constants of 1–5 $\mu M^{-1}\ s^{-1}$. The distal side is shaped by the side chains of Val7(B6), Phe10(B9), and Thr48(E11). The tunnel is hydrophobic, with a diameter around 6 Å in the thinnest part near Leu86(G12). The mutations of Leu86 to Ala, Phe, and Trp affect the size and shape of the tunnel, clearly affecting ligand migration, showing how the changes in the structure can modulate the association rates of gaseous ligands (Pesce et al., 2011; Salter et al., 2012, 2008). In this case, molecular simulations have contributed to understand the critical role of Leu86 in ligand migration. Thus, they showed a very small barrier for O_2 migration toward the active site at a distance of ~9 Å from the Fe, which corresponds to the vicinity of Leu86, but the main barrier (~5.5 kcal mol^{-1}) is found when the ligand passes close to PheB9, TyrB10, and ValB6. However, when Leu86 is replaced by Phe or Trp, the long channel is partially blocked (Fig. 4), explaining the decrease in ligand association constants and the increased geminate rebinding (Pesce et al., 2011).

Finally, the side chain of Gln44(E7) was also found to populate an "out" conformation, where it moves away from the distal cavity toward the solvent. This might suggest that ligands could enter the protein through a second path, which could work as a gate as in the case of HisE7 in Mb. However, mutation of Gln44(E7) to smaller residues does not affect the binding of small ligands, whereas mutations to bulkier residues trigger a small decrease in the association rate constant. This indicates that the E7 gate is not important in CerHb, which reinforces the relevance of the long apolar tunnel in ligand migration (Pesce et al., 2011).

2.4 Truncated Hbs

A representative member of this class of globin is the truncated HbN (trHbN) from *M. tuberculosis*, which plays an important role in the defense mechanism against nitric oxide generated by macrophages due to its NOD activity (Lama, Pawaria, & Dikshit, 2006; Pathania, Navani, Gardner, Gardner, & Dikshit, 2002). The reduced size of trHbs is reflected in the

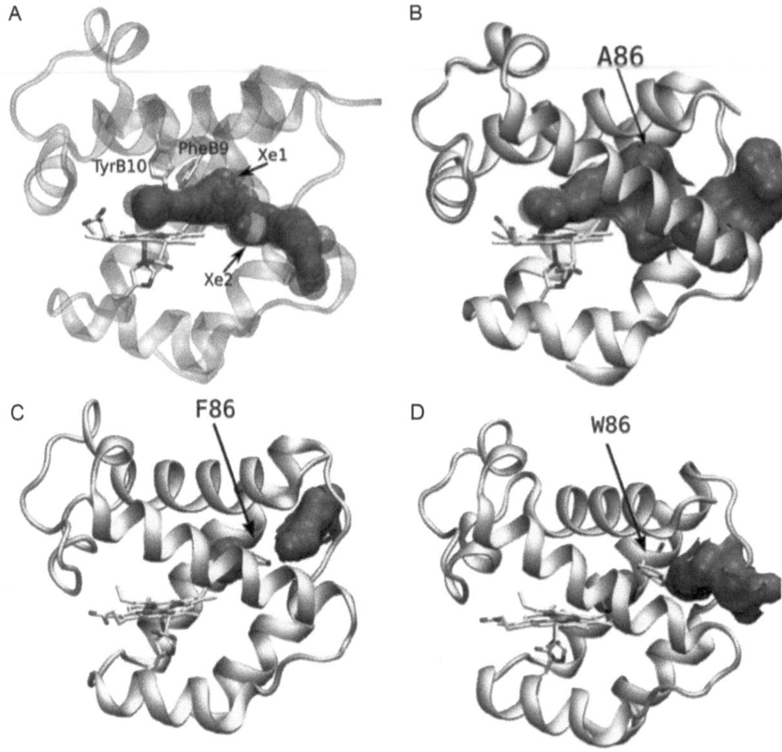

Fig. 4 Isocontour energy plots associated with the ligand migration pathway in wild type, Leu86A, Leu86F, and Leu86W mutants of CerHb (*panels A, B, C,* and *D*, respectively). The two Xe sites (Xe1 and Xe2) found in the crystal structure are depicted in panel (A), in *green* (*light gray* in the print version) *spheres. Reprinted from Pesce, A., Nardini, M., Dewilde, S., Capece, L., Martí, M. A., Congia, S., Salter, M. S., Blouin, G. C., Estrin, D. A., Ascenzi, P., Moens, L., Bolognesi, M., & Olson, J. S. (2011). Ligand migration in the apolar tunnel of Cerebratulus lacteus mini-hemoglobin.* Journal of Biological Chemistry, 286, *5347–5358.*

alteration of the protein fold, which is characterized by a 2-over-2 helical motif. Nevertheless, despite the smaller size relative to Mb, up to three different tunnels have been identified in trHbs: (i) the long tunnel (LT) (ii) and two short tunnels (STG8 and short tunnel E7 (STE7)).

In the specific case of trHbN, only the LT and STG8 tunnels appear to be well defined in the interior of the protein, allowing the exchange of ligands with the bulk solvent (Fig. 5). In particular, theoretical studies have suggested a dual-path ligand-induced dynamical mechanism that regulates the access of O_2 and NO to the heme cavity and to accomplish the conversion of NO to nitrate anion (Bidon-Chanal et al., 2006; Crespo et al., 2005). According to this mechanism, in the deoxygenated state of trHbN, O_2

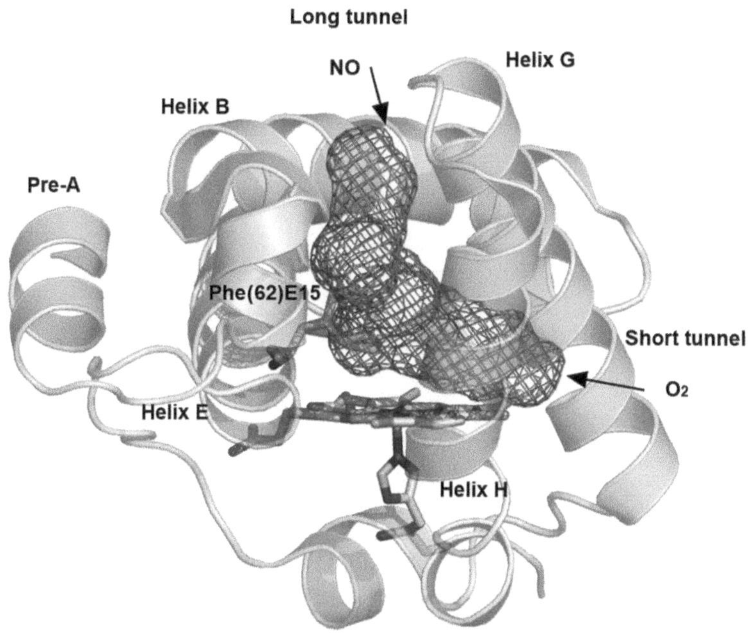

Fig. 5 Representation (based on the X-ray crystallographic structure 1IDR) of the *M. tuberculosis* trHbN and the access routes of O_2 and NO in the dual-path ligand-modulated mechanism proposed for this protein. The gating residue PheE15 (residue 62) is shown in the two conformations found in the X-ray structure as *sticks*. *Reprinted from Oliveira, A., Singh, S., Bidon-Chanal, A., Forti, F., Martí, M. A., Boechi, L., Estrin, D. A., Dikshit, K. L., & Luque, F. J. (2012). Role of PheE15 gate in ligand entry and nitric oxide detoxification function of Mycobacterium tuberculosis truncated hemoglobin N.* PloS One, *7, e49291.*

migrates to the heme active through the short tunnel. Binding to the heme Fe then triggers a local conformational change in the pair of residues formed by TyrB10 and GlnE11, which are hydrogen bonded. This conformational change favors the collision of GlnE11 with the side change of PheE15, a gating residues that regulates migration through the LT channel, thus facilitating the transition between closed and open forms of this tunnel branch. Furthermore, binding of oxygen promotes a global change in the dynamics of the protein backbone, which tends to enhance the relative motion of helices B and E (ie, those that primarily shape the LT channel), thus facilitating the transitions of the gating residue due to a reduced friction between the two helices. The net effect is that nitric oxide, which likely accumulates at the hydrophobic entrance of the LT channel, can then migrate to the heme cavity and react with the heme-bound O_2, giving rise to the nitrate anion.

This mechanism is supported by a number of experimental findings. For instance, the change in the dynamical motion of the protein backbone triggered upon binding of O_2 to the heme agrees with the large-scale dynamical change observed upon ligand binding to trHbN (Mukai, Ouellet, Guertin, & Yeh, 2004). Furthermore, the molecular switch formed by the pair TyrB10–GlnE11 is greatly affected by TyrB10→Phe and GlnE11→Ala mutations, as expected from the inability to exert a mechanical pressure on the PheE15 gating residue, which would preferentially remain in the closed state (Bidon-Chanal, Martí, Estrín, & Luque, 2007). On the other hand, it has been shown that deletion of Pre-A region from *M. tuberculosis* trHbN drastically reduces its NO-scavenging activity, whereas its insertion at the N-terminus of *M. smegmatis* trHbN, which lacks Pre-A segment, improved its NO dioxygenase activity. These changes were attributed to the changes in protein dynamics arising from excision of Pre-A motif in trHbN, which cause the gate of the tunnel long branch to be trapped into a closed conformation, thus impeding migration of diatomic ligands toward the heme active site (Lama et al., 2009), although an indirect effect related to preventing dimerization of HbN has also been proposed to explain the important effect of Pre-A region on the NOD activity (Pesce et al., 2016). Finally, the most direct evidence comes from mutagenesis studies in which the PheE15 gating residue was mutated to alanine, isoleucine, tyrosine, and tryptophan (Oliveira et al., 2012). While mutants have similar O_2/CO association kinetics, they display significant reduction in their NOD function, an effect directly attributed to the disruption in the accessibility through the LT as found from the analysis of molecular dynamics simulations (Fig. 6).

In contrast with the topological features of trHbN, our studies on trHbO from different organisms, such as *M. tuberculosis*, *Thermobifida fusca*, *Bacillus subtilis*, and *Campylobacter jejuni*, have consistently shown that the STG8 tunnel branch is blocked by the presence of TrpG8 (Boechi, Máñez, Luque, Martí, & Estrín, 2010; Boechi et al., 2008). Therefore, ligands should migrate through the LT tunnel or use an additional path similar to that found in the Mb, that is, the STE7. Furthermore, besides the gating residue PheE15, position E11 also plays a crucial role for the opening of the LT branch. Thus, in *B. subtilis* trHbO, this position is occupied by a Gln residue, which can make a hydrogen bond with TyrB10, and hence assist the opening of the LT tunnel, as was suggested for *M. tuberculosis* trHbN (see earlier). However, in *M. tuberculosis* trHbO, position E11 is occupied by a Leu residue, and then the LT branch tends to remain closed, which makes STE7 to be the major pathway for the migration of the ligands. This simple mutation

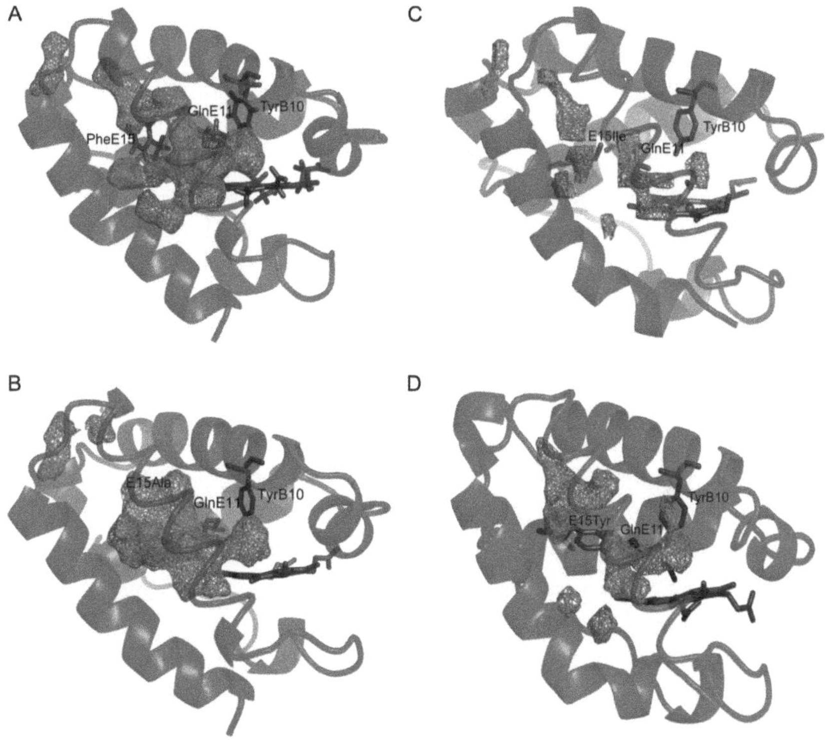

Fig. 6 Representation of the accessible volume in (A) *M. tuberculosis* trHbN and the mutants, (B) PheE15→Ala, (C) PheE15→Ile, and (D) PheE15→Tyr. In all mutants the LT tunnel is disrupted at distinct sites, thus preventing the access of NO to the heme cavity. *Reprinted from Oliveira, A., Singh, S., Bidon-Chanal, A., Forti, F., Martí, M. A., Boechi, L., Estrin, D. A., Dikshit, K. L., & Luque, F. J. (2012). Role of PheE15 gate in ligand entry and nitric oxide detoxification function of Mycobacterium tuberculosis truncated hemoglobin N.* PloS One, *7, e49291.*

is proposed to be the reason that explains why *B. subtilis* trHbO allows a faster uptake of ligand (100-fold) than in *M. tuberculosis* trHbO.

As a last remark, let us briefly mention a recent study that addressed the conversion of truncated Hbs N and O, specifically regarding the role played by residue G8 in determining the structural shape of accessible ligand migration routes in these proteins (Boron, Bustamante, et al., 2015). In this work, by combining mutagenesis studies that involved the change of ValG8 present in trHbN to both Phe and Trp (present in trHbO) with kinetic measurements and molecular dynamics simulations, it was shown that these mutations altered the tunnel topologies as well as the ability of the incoming ligand to displace retained water molecules in the heme cavity. Remarkably, trHbN acquired ligand migration rates comparable to those observed for trHbO. Therefore, this finding confirms the crucial role played by the inner tunnel architecture in regulating ligand migration in these proteins.

2.5 Protoglobin

A different globin subgroup includes Pgb, which is the first single-domain globin-coupled sensor-related globin found in *Archaea* (Freitas et al., 2004). Pgb is bigger than Mb, as its contain around 190 residues, and is able to reversibly bind O_2, CO, and NO. The in vivo function remains, however, unknown yet. A number of peculiar structural traits can be identified from the X-ray structure of *Methanosarcina acetivorans* Pgb, such as the presence of nine helices (a Pre-A helix named Z was clearly found in the crystallographic structure), and a large structural distortion of the heme moiety, which is completely occluded in the protein matrix by the CE and FG loops and a 20-residue long N-terminal loop (Nardini et al., 2008). Finally, as many other globin-coupled sensor proteins, Pgb behaves as a dimer in solution, with an interface contributed mostly by residues belonging to helices G and H, which build an intermolecular four-helix bundle.

Ligands may access the heme cavity through two tunnels mainly defined by helices B/G (tunnel 1) and B/E (tunnel 2), which form an apolar V-shaped channel (Fig. 7). This topology is a unique feature of Pgb, as it differs from the system of internal cavities found in other globins. These structural features, together with an unusually low O_2 dissociation rate

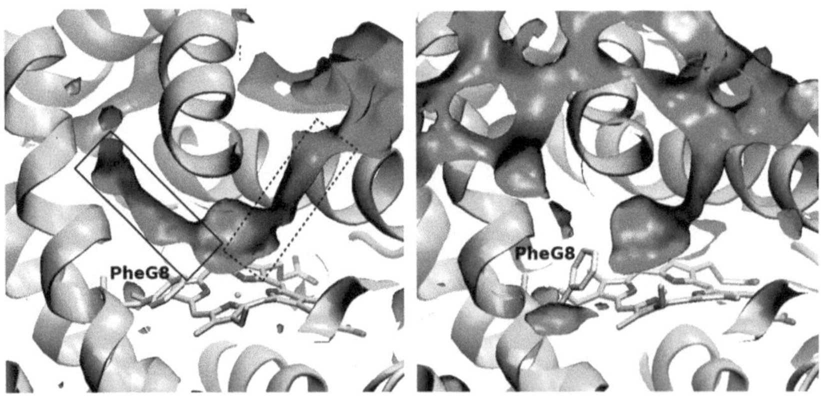

Fig. 7 Representation of tunnels 1 and 2 in Pgb. Whereas tunnel 2 (enclosed in *dashed line*) is fully accessible, tunnel 1 (enclosed in *solid line*) may populate both open (*left*) and closed (*right*) states depending on the conformation of Phe(145)G8. *Reprinted from Forti, F., Boechi, L., Bikiel, D., Martí, M. A., Nardini, M., Bolognesi, M., Viappiani, C., Estrín, D. A., & Luque, F. J. (2011). Ligand migration in Methanosarcina acetivorans protoglobin: Effects of ligand binding and dimeric assembly.* Journal of Physical Chemistry B, 115, *13771–13780.*

constant, make *Ma*Pgb a relevant protein for gaining insight into the ligand migration from molecular simulations (Forti et al., 2011).

The results of these studies have revealed a number of key events. First, the accessibility of ligands through tunnel 1 is regulated by Phe(145)G8, which is found in two main conformations along the molecular simulations (Fig. 7). In one case (ie, the open conformation), access through tunnel 1 is not hindered, but in the other (ie, the closed state), the phenyl ring protrudes into the tunnel and hampers the migration of ligands. Second, the relative population of the two conformations depends on the assembly state of the protein and on the presence of the ligand in the distal cavity. Thus, dimerization of the protein increases the population of the open state, and this trend is significantly enhanced upon binding of O_2 to the distal cavity. Third, the enhanced conformational fluctuations of helix B triggered upon O_2 binding also facilitates ligand migration through tunnel 1. Finally, tunnel 2 is found to be accessible irrespective of the presence of heme-bound ligand and of the monomeric/dimeric state of the protein.

Overall, these findings suggest that Pgb might participate in a ligand-controlled bimolecular chemical process. Accordingly, a small ligand could access the heme cavity through tunnel 2, which seems to be fully accessible, while binding to the heme would result in the opening of the second tunnel. The physiological implications of these findings remain unknown. However, they agree with experimental data collected from rebinding experiments, which indicate the existence of a ligand-dependent equilibrium between two conformational species corresponding to fast and slow rebinding processes, and ligation favors the fast rebinding phase.

As a last remark, it is worth noting the striking similarity between trHbN and Pgb, as far as these two distinct globins exhibit a dual-path ligand-mediated migration mechanism, which in the two cases requires the binding of a ligand to the heme to trigger the opening of the second tunnel, making it accessible for the migration of a second ligand. Furthermore, some common regulatory elements can also be recognized, such as the presence of a gating residue, the change in conformational motions of the proteins, and the influence of dimerization on the accessibility of the tunnels.

3. THE COORDINATION OF THE HEME GROUP AND ITS ROLE IN LIGAND BINDING

3.1 Neuroglobin

Most of the globin functions rely on the binding of external ligands to the distal position of the heme group. Often the distal site is occupied by a protein residue belonging to the E helix, generally a His, forming a bis-histidyl hexacoordinated (6c) heme. Therefore, the 6c ⇆ 5c equilibrium provides an additional mechanism to regulate ligand binding, as will be shown here for the particular case of Ngb.

Ngb is an ancient globin present in the nervous system of vertebrates, with a high level of expression in highly oxygen-consuming tissues such as the retina (Brunori & Vallone, 2007; Burmester, Weich, Reinhardt, & Hankeln, 2000), but its in vivo function is still unclear (Burmester & Hankeln, 2009). The X-ray structure shows a monomeric protein with the classical 3-over-3 globin fold, wherein HisE7 occupies the distal coordination site forming an hexacoordinated heme. A peculiarity trait of Ngb is the presence of a large disordered region between the C-terminus of the B helix and the N-terminal side of the E helix. In contrast to Mb, which displays two short helixes (C and D), in Ngb, this region is almost unstructured, showing only a short piece of the C helix. Finally, the cavity found in Ngb is huge (~300 $Å^3$) and involves the four ligand-docking sites found in Mb. The changes in the coordination state, as well as the redox state of disulfide bonds can affect the protein dynamics and therefore the size of inner cavities (Bocahut, Bernad, Sebban, & Sacquin-Mora, 2009). Indeed, the 6c ⇆ 5c equilibrium has shown to be modulated by the formation of a disulfide bridge, which displaces the equilibrium toward the 5c state, increasing the O_2 affinity by a factor of 10 (Hamdane et al., 2003).

With the aim of understanding the 6c ⇆ 5c transition, theoretical studies were performed for Ngb in its 6c, 5c, and oxygenated forms (Nadra, Martí, Pesce, Bolognesi, & Estrín, 2008), for ferrous Ngb in the 6c, and 5c, as well as in the presence of disulfide bridges. Most of the structural differences were found in the CD region, and particularly formation of the disulfide bridge between residues CysCD7 and CysD5 led to a stabilization of the pentacoordinated species, increasing the barrier for hexacoordination. These findings support the existence of a molecular mechanism whereby O_2 would be released under hypoxic conditions, which in turn would suggest that Ngb may be involved in O_2 storage in the brain.

The influence of pressure on the 6c ⇆ 5c equilibrium has also been examined, since it was experimentally found that high pressure favored the formation of the 6c state (Capece et al., 2009). Molecular simulations showed that high-pressure conditions reduce the flexibility of the protein structure, where the most mobile regions are particularly affected, as the CD region in the case of Ngb. Furthermore, free-energy calculations indicated that the enhanced pressure-induced rigidity modifies the 6c ⇆ 5c equilibrium, favoring the 6c state. These results reinforce the importance of protein flexibility in the hexacoordination process.

The dynamical properties of Ngb and Mb, taken as representative members of 6c and 5c globins, have been recently compared (Boron, Capece, et al., 2015). The results obtained from molecular dynamics simulations indicate that the CD region of Ngb displays a much larger flexibility than in Mb, reinforcing the importance of the CD region flexibility in driving the 6c ⇆ 5c equilibrium. To further confirm this observation, chimeric proteins were constructed, where the corresponding CD regions of Ngb and Mb were exchanged and examined by combining experimental and theoretical techniques (Fig. 8). Interestingly, it was observed that chimeric Mb retained the flexibility of the CD region of Ngb, and importantly it was able to partially

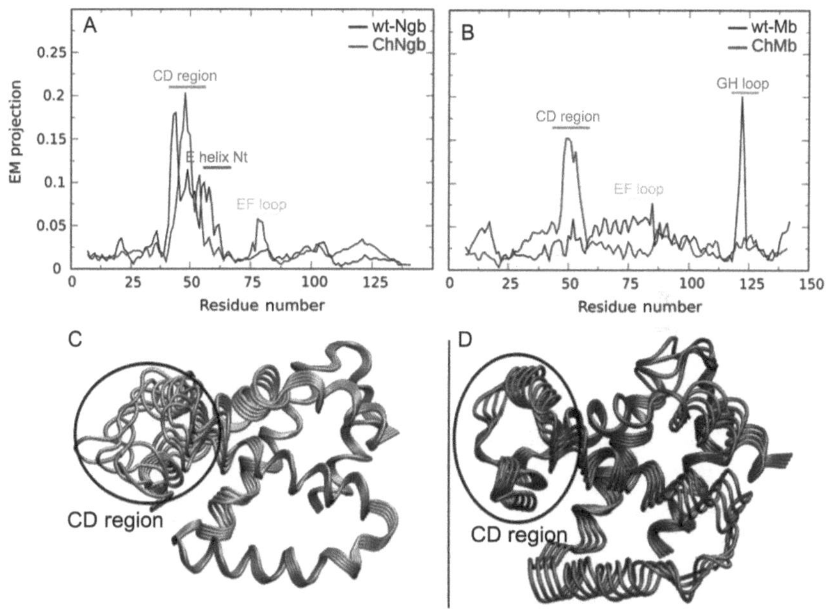

Fig. 8 Comparative analysis of the 6c ⇆ 5c transition in Ngb and Mb. Panels (A and B) show the projection of the transition essential mode obtained by combining the molecular dynamics trajectories of the 5c and 6c states, in both the wild-type (wt) and chimeric proteins (Ch). Panels (C and D) show superimposed snapshots of the animation of the transition essential mode for wt-Ngb and wt-Mb, respectively. *Reprinted from Boron, I., Capece, L., Pennacchietti, F., Wetzler, D. E., Bruno, S., Abbruzzetti, S., Chisari, L., Luque, F. J., Viappiani, C., Martí, M. A., Estrín, D. A., & Nadra, A. D. (2015). Engineered chimeras reveal the structural basis of hexacoordination in globins: A case study of neuroglobin and myoglobin.* Biochimica et Biophysica Acta, 1850, *169–177.*

form the 6c state. These findings confirm the importance of the CD region in regulating the hexacoordination equilibrium, and eventually the affinity of ligands in Ngb (Fig. 8).

4. CONCLUSIONS

The preceding cases illustrate the strong connection between the specific sequence of residues and the topological features of the inner cavities and tunnels that assist exchange of ligands to/from the heme cavity in globins. Furthermore, they also provide a sound basis to disclose the relationships between structure, dynamics, and function. Indeed, it can be argued that a proper understanding of ligand migration cannot be achieved without a deep knowledge of the protein dynamics. The representative examples examined here illustrate the success of current experimental and computational techniques in deciphering the molecular mechanisms of the migration processes.

The major challenge is to move from the qualitative description gained from these studies to a quantitative determination of microscopic rates and energetic profiles, which may lead to estimate binding affinities and association/dissociation rate constants fully comparable with the values obtained from experimental measurements. Clearly, a synergistic approach that combines structural information about internal-docking sites with chemically accurate estimates of the free-energy landscape for ligand migration will be extremely valuable.

The increase in computer power seen in the last years facilitates the analysis of multiple independent molecular dynamics simulations of ligand migrating through the protein, enabling an exhaustive sampling of the different migration pathways, and more importantly the derivation of the free-energy landscape. Preliminary studies support the feasibility of these computational strategies (Anselmi, DiNola, & Amadei, 2011; Cazade et al., 2015; Mishra & Meuwly, 2010; Wang, Best, & Blumberger, 2011). These advances, in conjunction with the development of enhanced sampling techniques and the development of refined force fields, or even the inclusion of quantum mechanical calculations, will facilitate the transition to a quantitative description of the kinetic mechanisms of ligand (un)binding. In turn, this will contribute to gain a comprehensive understanding of the structure–function relationships in globins but will also afford the basis for chemical engineering and biotechnological applications of these proteins.

ACKNOWLEDGMENTS

We thank the financial support from Ministerio de Economía y Competitividad (MINECO; SAF2014-57094-R) and the Generalitat de Catalunya (GC; 2014SGR1189). We are grateful to the Consorci de Serveis Universitaris de Catalunya for computational resources. F.J.L. acknowledges the support from ICREA Academia. Fellowship (FPDI-2013-15572) from MINECO to C.E. is gratefully acknowledged. D.E. and L.C. thank the University of Buenos Aires, the Agencia de Producción Científica y Tecnológica (Argentina) and CONICET for financial support.

REFERENCES

Abbruzzetti, S., Spyrakis, F., Bidon-Chanal, A., Luque, F. J., & Viappiani, C. (2013). Ligand migration through hemeprotein cavities: Insights from laser flash photolysis and molecular dynamics simulations. *Physical Chemistry Chemical Physics*, *15*, 10686–10701.

Anselmi, M., DiNola, A., & Amadei, A. (2011). Kinetics of carbon monoxide migration and binding in solvated neuroglobin as revealed by molecular dynamics simulations and quantum mechanical calculations. *Journal of Physical Chemistry B*, *115*, 2436–2446.

Arroyo-Mañez, P., Bikiel, D. E., Boechi, L., Capece, L., Di Lella, S., Estrín, D. A., et al. (2011). Protein dynamics and ligand migration interplay as studied by computer simulation. *Biochimica et Biophysica Acta*, *1814*, 913–1106.

Banushkina, P., & Meuwly, M. (2005). Free-energy barriers in MbCO rebinding. *Journal of Physical Chemistry B*, *109*, 16911–16917.

Bidon-Chanal, A., Martí, M. A., Crespo, A., Milani, M., Orozco, M., Bolognesi, M., et al. (2006). Ligand-induced dynamical regulation of NO conversion in Mycobacterium tuberculosis truncated hemoglobin-N. *Proteins*, *64*, 457–464.

Bidon-Chanal, A., Martí, M. A., Estrín, D. A., & Luque, F. J. (2007). Dynamical regulation of ligand migration by a gate-opening molecular switch in truncated hemoglobin-N from Mycobacterium tuberculosis. *Journal of the American Chemical Society*, *129*, 6782–6788.

Birukou, I., Schweers, R. L., & Olson, J. S. (2010). Distal histidine stabilizes bound O_2 and acts as a gate for ligand entry in both subunits of adult human hemoglobin. *Journal of Biological Chemistry*, *285*, 8840–8854.

Birukou, I., Soman, J., & Olson, J. S. (2011). Blocking the gate to ligand entry in human hemoglobin. *Journal of Biological Chemistry*, *286*, 10515–10529.

Bocahut, A., Bernad, S., Sebban, P., & Sacquin-Mora, S. (2009). Relating the diffusion of small ligands in human neuroglobin to its structural and mechanical properties. *The Journal of Physical Chemistry. B*, *113*, 16257–16267.

Boechi, L., Arrar, M., Martí, M. A., Olson, J. S., Roitberg, A. E., & Estrín, D. A. (2013). Hydrophobic effect drives oxygen uptake in myoglobin. *Journal of Biological Chemistry*, *288*, 6754–6762.

Boechi, L., Mañez, P. A., Luque, F. J., Martí, A. M., & Estrín, D. A. (2010). Unraveling the molecular basis for ligand binding in truncated hemoglobins: The trHbO Bacillus subtilis case. *Proteins*, *78*, 962–970.

Boechi, L., Martí, M. A., Milani, M., Bolognesi, M., Luque, F. J., & Estrín, D. A. (2008). Structural determinants of ligand migration in Mycobacterium tuberculosis truncated hemoglobin O. *Proteins*, *73*, 372–379.

Bolognesi, M., Bordo, D., Rizzi, M., Tarricone, C., & Ascenzi, P. (1997). Nonvertebrate hemoglobins: Structural bases for reactivity. *Progress in Biophysics and Molecular Biology*, *68*, 29–68.

Boron, I., Bustamante, J. P., Davidge, K. S., Singh, S., Bowman, L. A., Tinajero-Trejo, M., et al. (2015). Ligand uptake in Mycobacterium tuberculosis truncated hemoglobins is controlled by both internal tunnels and active site water molecules. *F1000 Research*, *4*, 22.

Boron, I., Capece, L., Pennacchietti, F., Wetzler, D. E., Bruno, S., Abbruzzetti, S., et al. (2015). Engineered chimeras reveal the structural basis of hexacoordination in globins: A case study of neuroglobin and myoglobin. *Biochimica et Biophysica Acta*, *1850*, 169–177.

Bossa, C., Amadei, A., Daidone, I., Anselmi, M., Vallone, B., Brunori, M., et al. (2005). Molecular dynamics simulation of sperm whale myoglobin: Effects of mutations and trapped CO on the structure and dynamics of cavities. *Biophysical Journal*, *89*, 465–474.

Bossa, C., Anselmi, M., & Di Nola, A. (2004). Extended molecular dynamics simulation of the carbon monoxide migration in sperm whale myoglobin. *Biophysical Journal*, *86*, 3855–3862.

Brunori, M., & Vallone, B. (2007). Neuroglobin, seven years after. *Cellular and Molecular Life Sciences*, *64*, 1259–1268.

Burmester, T., & Hankeln, T. (2009). What is the function of neuroglobin? *Journal of Experimental Biology*, *212*, 1423–1428.

Burmester, T., Weich, B., Reinhardt, S., & Hankeln, T. (2000). A vertebrate globin expressed in the brain. *Nature*, *407*, 520–523.

Capece, L., Boechi, L., Perissinotti, L. L., Arroyo-Máñez, P., Bikiel, D. E., Smulevich, G., et al. (2013). Small ligand-globin interactions: Reviewing lessons derived from computer simulation. *Biochimica et Biophysica Acta*, *1834*, 1722–1738.

Capece, L., Martí, M. A., Bidon-Chanal, A., Nadra, A., Luque, F. J., & Estrín, D. A. (2009). High pressure reveals structural determinants for globin hexacoordination: Neuroglobin and myoglobin cases. *Proteins*, *75*, 885–894.

Cazade, P. A., Zheng, W., Prada-Gracia, D., Berezovska, G., Rao, F., Clementi, C., et al. (2015). A comparative analysis of clustering algorithms: O_2 migration in truncated hemoglobin I from transition networks. *Journal of Chemical Physics*, *142*, 025103.

Cho, H. S., Dashdorj, N., Schotte, F., Graber, T., Henning, R., & Anfinrud, P. (2010). Protein structural dynamics in solution unveiled via 100-ps time-resolved X-ray scattering. *Proceedings of the National Academy of Sciences of the United States of America*, *107*, 7281–7728.

Cohen, J., Arkhipov, A., Braun, R., & Schulten, K. (2006). Imaging the migration pathways for O_2, CO, NO, and Xe inside myoglobin. *Biophysical Journal*, *91*, 1844–1857.

Crespo, A., Martí, M. A., Kalko, S. G., Morreale, A., Orozco, M., Gelpí, J. L., et al. (2005). Theoretical study of the truncated hemoglobin HbN: Exploring the molecular basis of the NO detoxification mechanism. *Journal of the American Chemical Society*, *127*, 4433–4444.

Forti, F., Boechi, L., Bikiel, D., Martí, M. A., Nardini, M., Bolognesi, M., et al. (2011). Ligand migration in Methanosarcina acetivorans protoglobin: Effects of ligand binding and dimeric assembly. *Journal of Physical Chemistry B*, *115*, 13771–13780.

Frauenfelder, H., McMahon, B. H., & Fenimore, P. W. (2003). Myoglobin: The hydrogen atom of biology and a paradigm of complexity. *Proceedings of the National Academy of Sciences of the United States of America*, *100*, 8615–8617.

Freitas, T. A. K., Hou, S., Dioum, E. M., Saito, J. A., Newhouse, J., González, G., et al. (2004). Ancestral hemoglobins in Archaea. *Proceedings of the National Academy of Sciences of the United States of America*, *101*, 6675–6680.

Geuens, E., Dewilde, S., Hoogewijs, D., Pesce, A., Nienhaus, K., Nienhaus, G. U., et al. (2004). Nerve globins in invertebrates. *IUBMB Life*, *56*, 653–656.

Hamdane, D., Kiger, L., Dewilde, S., Green, B. N., Pesce, A., Uzan, J., et al. (2003). The redox state of the cell regulates the ligand binding affinity of human neuroglobin and cytoglobin. *Journal of Biological Chemistry*, *278*, 51713–51721.

Kendrew, J. C., Bodo, G., Dintzis, H. M., Parrish, R. G., Wyckoff, H., & Phillips, D. C. (1958). A three-dimensional model of the myoglobin molecule obtained by X-ray analysis. *Nature, 181*, 662–666.

Laberge, M., & Yonetani, T. (2007). Common dynamics of globin family proteins. *IUBMB Life, 59*, 528–534.

Lama, A., Pawaria, S., Bidon-Chanal, A., Anand, A., Gelpí, J. L., Arya, S., et al. (2009). Role of pre-A motif in nitric oxide scavenging by truncated hemoglobin, HbN, of Mycobacterium tuberculosis. *Journal of Biological Chemistry, 284*, 14457–14468.

Lama, A., Pawaria, S., & Dikshit, K. L. (2006). Oxygen binding and NO scavenging properties of truncated hemoglobin, HbN, of Mycobacterium smegmatis. *FEBS Letters, 580*, 4031–4041.

Lucas, F., & Guallar, V. (2012). An atomistic view on human hemoglobin carbon monoxide migration processes. *Biophysical Journal, 102*, 887–896.

Martí, M. A., Bikiel, D. E., Crespo, A., Nardini, M., Bolognesi, M., & Estrín, D. A. (2006). Two distinct heme distal site states define Cerebratulus lacteus mini-hemoglobin oxygen affinity. *Proteins, 62*, 641–648.

Mishra, S., & Meuwly, M. (2010). Quantitative analysis of ligand migration from transition networks. *Biophysical Journal, 99*, 3969–3978.

Mouawad, L., Marechal, J.-D., & Perahia, D. (2005). Internal cavities and ligand passageways in human hemoglobin characterized by molecular dynamics simulations. *Biochimica et Biophysica Acta, 1724*, 385–393.

Mukai, M., Ouellet, Y., Guertin, M., & Yeh, S.-R. (2004). NO binding induced conformational changes in a truncated hemoglobin from Mycobacterium tuberculosis. *Biochemistry, 43*, 2764–2770.

Nadra, A. D., Martí, M. A., Pesce, A., Bolognesi, M., & Estrín, D. A. (2008). Exploring the molecular basis of heme coordination in human neuroglobin. *Proteins, 71*, 695–705.

Nardini, M., Thijs, L., Saito, J. A., Dewilde, S., Alam, M., Ascenzi, P., et al. (2008). Archaeal protoglobin structure indicates new ligand diffusion paths and modulation of haem-reactivity. *EMBO Reports, 9*, 157–163.

Nutt, D. R., & Meuwly, M. (2004). CO migration in native and mutant myoglobin: Atomistic simulations for the understanding of protein function. *Proceedings of the National Academy of Sciences of the United States of America, 101*, 5998–6002.

Oliveira, A., Singh, S., Bidon-Chanal, A., Forti, F., Martí, M. A., Boechi, L., et al. (2012). Role of PheE15 gate in ligand entry and nitric oxide detoxification function of Mycobacterium tuberculosis truncated hemoglobin N. *PloS One*, 7, e49291.

Olson, J. S., Soman, J., & Phillips, G. N., Jr. (2007). Ligand pathways in myoglobin: A review of Trp cavity mutations. *IUBMB Life, 59*, 552–562.

Pathania, R., Navani, N. K., Gardner, A. M., Gardner, P. R., & Dikshit, K. L. (2002). Nitric oxide scavenging and detoxification by the Mycobacterium tuberculosis haemoglobin, HbN in Escherichia coli. *Molecular Microbiology, 45*, 1303–1314.

Pesce, A., Bustamante, J. P., Bidon-Chanal, A., Boechi, L., Estrin, D. A., Luque, F. J., et al. (2016). The N-terminal pre-A region of Mycobacterium tuberculosis 2/2HbN promotes NO-dioxygenase activity. *FEBS Journal, 283*, 305–322.

Pesce, A., Dewilde, S., Kiger, L., Milani, M., Ascenzi, P., Marden, M. C., et al. (2001). Very high resolution structure of a trematode hemoglobin displaying a TyrB10-TyrE7 heme distal residue pair and high oxygen affinity. *Journal of Molecular Biology, 309*, 1153–1164.

Pesce, A., Milani, M., Nardini, M., & Bolognesi, M. (2008). Mapping heme-ligand tunnels in group I truncated (2/2) hemoglobins. *Methods in Enzymology, 436*, 303–315.

Pesce, A., Nardini, M., Ascenzi, P., Geuens, E., Dewilde, S., Moens, L., et al. (2004). Thr-E11 regulates O_2 affinity in Cerebratulus lacteus mini-hemoglobin. *Journal of Biological Chemistry, 279*, 33662–33672.

Pesce, A., Nardini, M., Dewilde, S., Capece, L., Martí, M. A., Congia, S., et al. (2011). Ligand migration in the apolar tunnel of Cerebratulus lacteus mini-hemoglobin. *Journal of Biological Chemistry*, *286*, 5347–5358.

Ruscio, J. Z., Kumar, D., Shukla, M., Prisant, M. G., Murali, T. M., & Onufriev, A. V. (2008). Atomic level computational identification of ligand migration pathways between solvent and binding site in myoglobin. *Proceedings of the National Academy of Sciences of the United States of America*, *105*, 9204–9920.

Salter, M. D., Blouin, G. C., Soman, J., Singleton, E. W., Dewilde, S., Moens, L., et al. (2012). Determination of ligand pathways in globins: Apolar tunnels versus polar gates. *Journal of Biological Chemistry*, *287*, 33163–33178.

Salter, M. D., Nienhaus, K., Nienhaus, G. U., Dewilde, S., Moens, L., Pesce, A., et al. (2008). The apolar channel in Cerebratulus lacteus hemoglobin is the route for O_2 entry and exit. *Journal of Biological Chemistry*, *283*, 35689–35702.

Sanctis, D., Dewilde, S., Pesce, A., Moens, L., Ascenzi, P., Hankeln, T., et al. (2004). Mapping protein matrix cavities in human cytoglobin through Xe atom binding. *Biochemical and Biophysical Research Communications*, *316*, 1217–1221.

Schmidt, M., Nienhaus, K., Pahl, R., Krasselt, A., Anderson, S., Parak, F., et al. (2005). Ligand migration pathway and protein dynamics in myoglobin: A time-resolved crystallographic study on L29W MbCO. *Proceedings of the National Academy of Sciences of the United States of America*, *102*, 11704–11709.

Schotte, F., Lim, M., Jackson, T. A., Smirnov, A. V., Soman, J., Olson, J. S., et al. (2003). Watching a protein as it functions with 150-ps time-resolved x-ray crystallography. *Science*, *300*, 1944–1947.

Scott, E. E., Gibson, O. H., & Olson, J. S. (2001). Mapping the pathways for O_2 entry into and exit from myoglobin. *Journal of Biological Chemistry*, *276*, 5177–5188.

Tejero, J., & Gladwin, M. T. (2014). The globin superfamily: Functions in nitric oxide formation and decay. *Biological Chemistry*, *395*, 631–639.

Tilton, R. F., Kuntz, I. D., & Petsko, G. A. (1984). Cavities in proteins: Structure of a metmyoglobin-xenon complex solved to 1.9 A. *Biochemistry*, *23*, 2849–2857.

Tomita, A., Sato, T., Ichiyanagi, K., Nozawa, S., Ichikawa, H., Chollet, M., et al. (2009). Visualizing breathing motion of internal cavities in concert with ligand migration in myoglobin. *Proceedings of the National Academy of Sciences of the United States of America*, *106*, 2612–2616.

Vandergon, T. L., Riggs, C. K., Gorr, T. A., Colacino, J. M., & Riggs, A. F. (1998). The mini-hemoglobins in neural and body wall tissue of the nemertean worm, Cerebratulus lacteus. *Journal of Biological Chemistry*, *273*, 16998–17011.

Vinogradov, S. N., & Moens, L. (2008). Diversity of globin function: Enzymatic, transport, storage, and sensing. *Journal of Biological Chemistry*, *283*, 8773–8777.

Wajcman, H., Kiger, L., & Marden, M. C. (2009). Structure and function evolution in the superfamily of globins. *Comptes Rendus Biologies*, *332*, 273–282.

Wang, P.-H., Best, R. B., & Blumberger, J. (2011). A microscopic model for gas diffusion dynamics in a [NiFe]-hydrogenase. *Physical Chemistry Chemical Physics*, *13*, 7708–7719.

Yeh, S.-R., Couture, M., Ouellet, Y., Guertin, M., & Rousseau, D. L. (2000). A cooperative oxygen binding hemoglobin from Mycobacterium tuberculosis. Stabilization of heme ligands by a distal tyrosine residue. *Journal of Biological Chemistry*, *275*, 1679–1684.

CHAPTER FOUR

Dynamic Reorganization and Enzymatic Remodeling of Type IV Collagen at Cell–Biomaterial Interface

N.M. Coelho[*,1], V. Llopis-Hernández[†,‡], M. Salmerón-Sánchez[‡], G. Altankov[*,§,2]

[*]Institute for Bioengineering of Catalonia (IBEC), Barcelona, Spain
[†]Center for Biomaterials and Tissue Engineering, Universidad Politécnica de Valencia, Valencia, Spain
[‡]School of Engineering, University of Glasgow, Glasgow, United Kingdom
[§]Institució Catalana de Recerca i Estudis Avançats (ICREA), Barcelona, Spain
[2]Corresponding author: e-mail address: altankov@abv.bg

Contents

1. Introduction 82
2. Material and Methods 85
   2.1 Model Surfaces 85
   2.2 Protein Adsorption 86
   2.3 Atomic Force Microscopy 87
   2.4 Cells 87
   2.5 Initial Cellular Interaction 87
   2.6 Immunofluorescence 88
   2.7 Visualization of Adsorbed Collagen IV 89
   2.8 Western Blot Analysis 89
   2.9 Collagen IV Degradation 90
3. Results and Discussion 91
   3.1 Remodeling of Col IV by Endothelial Cells on Bare NH_2 and CH_3 Surfaces 91
   3.2 Morphological Observations 91
   3.3 Reorganization and Enzymatic Remodeling of Collagen IV on Bare NH_2 and CH_3 Surfaces 92
   3.4 Quantitative Studies on Enzymatic Remodeling on NH_2 and CH_3 Surfaces 94
   3.5 Collagen IV Adsorption and Assembly on Mixed NH_2/CH_3 Surfaces 94
   3.6 HUVEC Interaction with Collagen IV on Mixed NH_2/CH_3 Surfaces 96
   3.7 Collagen IV Remodeling on Mixed NH_2/CH_3 SAMs 98
   3.8 Quantitative Studies on Enzymatic Collagen IV Remodeling 98
4. Conclusive Remarks 100
References 101

[1] Present address: Matrix Dynamics Group, Faculty of Dentistry, University of Toronto, Toronto, ON, Canada.

Advances in Protein Chemistry and Structural Biology, Volume 105
ISSN 1876-1623
http://dx.doi.org/10.1016/bs.apcsb.2016.06.001

Abstract

Vascular basement membrane remodeling involves assembly and degradation of its main constituents, type IV collagen (Col IV) and laminin, which is critical during development, angiogenesis, and tissue repair. Remodeling can also occur at cell–biomaterials interface altering significantly the biocompatibility of implants. Here we describe the fate of adsorbed Col IV in contact with endothelial cells adhering on positively charged NH_2 or hydrophobic CH_3 substrata, both based on self-assembly monolayers (SAMs) and studied alone or mixed in different proportions. AFM studies revealed distinct pattern of adsorbed Col IV, varying from single molecular deposition on pure NH_2 to network-like assembly on mixed SAMs, turning to big globular aggregates on bare CH_3. Human umbilical endothelial cells (HUVECs) interact better with Col IV adsorbed as single molecules on NH_2 surface and readily rearrange it in fibril-like pattern that coincide with secreted fibronectin fibrils. The cells show flattened morphology and well-developed focal adhesion complexes that are rich on phosphorylated FAK while expressing markedly low pericellular proteolytic activity. Conversely, on hydrophobic CH_3 substrata HUVECs showed abrogated spreading and FAK phosphorylation, combined with less reorganization of the aggregated Col IV and significantly increased proteolytic activity. The later involves both MMP-2 and MMP-9, as measured by zymography and FITC-Col IV release. The mixed SAMs support intermediate remodeling activity. Taken together these results show that chemical functionalization combined with Col IV pre-adsorption provides a tool for guiding the endothelial cells behavior and pericellular proteolytic activity, events that strongly affect the fate of cardiovascular implants.

1. INTRODUCTION

The behavior of cells within tissues is strongly dependent on the extracellular matrix (ECM)—a hierarchically organized structure consisting mostly of proteins and proteoglycans (Daley, Peters, & Larsen, 2008; Frantz, Stewart, & Weaver, 2010). It relates also to the basement membrane (BM)—a specialized type of ECM supporting highly organized tissues such as the endothelium, epithelia, and muscles. The BM contains various spatiotemporal cues that virtually impact all aspects of cell functionality, including cell growth, survival, migration, and differentiation (Aumailley & Gayraud, 1998; Goody & Henry, 2010; Hynes, 2009; Kalluri, 2003; Kolahi & Mofrad, 2010). The BM is highly dynamic structure that is continuously being remodeled by resident cells during development, homeostasis, and tissue repair. Dysregulated BM remodeling may result in pathologies such as fibrosis, inflammation, or tumor progression (Ala-aho & Kähäri, 2005; Daley et al., 2008; Kalluri, 2003; Larsen, Artym, Green, & Yamada, 2006; Wynn, 2008). BM remodeling comprises the production, deposition, and organization of BM molecules balanced by the degradation and remodeling of the existing BM (Shi, Harman, Fujiwara, & Sottile, 2010).

A large and growing body of evidences show that the cells need to accept distinct mechanical stimuli from the surrounding structures to strengthen their connections with the cytoskeleton, and thus responding to the mechanical properties of the environment (Geiger, Spatz, & Bershadsky, 2009; Janmey & McCulloch, 2007). As the stiffness of the surrounding ECM in vivo is in the same order of magnitude as cells, they tend to reorganize it mechanically in a way optimal for their functioning (Grinnell, 1986; Hubbell, 2003). Alternatively, cells can degrade the excess of ECM triggering the process of pericellular proteolysis (Cox & Erler, 2011; Lu, Takai, Weaver, & Werb, 2011). Pericellular proteolysis is a fundamental process in ECM remodeling and is critically involved in the control of ECM structure, composition, and also in the release of biologically active molecules (such as growth factors) that strongly impact the function of resident cells. Matrix metalloproteinases (MMPs) are a family of zinc-dependent, multidomain, endopeptidases that are characterized in part by a catalytic domain and a hemopexin domain. MMPs are typically soluble or membrane bound, and can collectively, cleave most if not all components of the ECM (Ala-aho & Kähäri, 2005; Page-McCaw, Ewald, & Werb, 2007; Sternlicht & Werb, 2001).

ECM undergo remodeling also in contact with biomaterials (Irvine et al., 2011; Llopis-Hernández, Rico, Ballester-Beltrán, Moratal, & Salmerón-Sánchez, 2011; Place, Evans, & Stevens, 2009) as the molecular events that take place at biomaterials interface mimic to a certain extent the natural cell–ECM interaction (Altankov et al., 2010; Ratner & Bryant, 2004; Stevens & George, 2005). Cell–materials interaction starts with the adsorption of matrix proteins from the surrounding medium (Grinnell & Feld, 1982) further recognized by integrins at distinct site of their molecule (eg, RGD, GFOGER, etc.) (García, 2005; Hynes, 2002). When integrins bind their ligand they cluster in focal adhesion complexes where specific bidirectional signaling converges with other signaling pathways (García, 2005; Hynes, 2002). It involves a common tyrosine phosphorylation mechanism (Geiger, Bershadsky, Pankov, & Yamada, 2001) that include focal adhesion kinase (FAK), Src, Cas, and paxillin within other important signaling molecules (Yamada & Geiger, 1997). On the level of cell–biomaterials interaction, however, depending on the conformation of adsorbed proteins, different cellular response may occur (Altankov & Groth, 1996; Keselowsky, Collard, & García, 2003, 2004), which influence strongly the biocompatibility of a material (Käpylä et al., 2000; Keresztes, Rouxhet, Remacle, & Dupont-Gillain, 2006).

This process of remodeling, however, may be hampered upon implantation of a material as exemplified by overaccumulation of ECM forming fibrous capsule around the implant (Stamenkovic, 2003; Thevenot, Hu, & Tang, 2008), or the implant failure because of the gap formation caused by missing ECM deposition (Daley et al., 2008). Collagens are the major component of this peri-implant tissue while the resident mesenchymal cells are generally involved in its synthesis (Abrahamsson et al., 2002; Moon, Berglundh, Abrahamsson, Linder, & Lindhe, 1999). Collagens play also an important role in the fate of cardiovascular implants, particularly type IV collagen (Col IV), the main structural constituent of the BM, including the vascular one. Recently, we have been exploring the possibility of combining chemical functionalization with Col IV preadsorption to engineer biomimetic interfaces that improve endothelization and vascularization of implants.

Our previous and more basic studies on the cell–biomaterials interaction (Altankov & Groth, 1994; Gustavsson et al., 2008; Maneva-Radicheva, Ebert, Dimoudis, & Altankov, 2008; Toromanov et al., 2015; Tzoneva, Groth, Altankov, & Paul, 2002) led us to hypothesize that tissue compatibility of materials depends on the allowance of cells to remodel adsorbed soluble matrix proteins such as fibronectin (FN), vitronectin, and fibrinogen, and to organize them in a provisional ECM (Altankov et al., 2010). We further anticipate that under specific conditions, other less-soluble ECM proteins like collagens may also associate with the biomaterial interface and the cellular response they elicit might be of significant bioengineering interest (Altankov et al., 2010). One such protein is Col IV (Coelho, Gonzalez-Garcia, Planell, Salmeron-Sanchez, & Altankov, 2010; Coelho, Gonzalez-Garcia, Salmeron-Sanchez, & Altankov, 2011a, 2011b; Coelho, Salmeron-Sanchez, & Altankov, 2013), which as the main structural component of the BM (Khoshnoodi, Pedchenko, & Hudson, 2008; Kühn, 1995; Timpl & Brown, 1996) plays crucial role in various physiological and pathological processes, including embryonic development, tissue remodeling, angiogenesis, and cancer (Kalluri, 2003; Khoshnoodi et al., 2008; Yurchenco & Schittny, 1990). Col IV is recognized by $\alpha1\beta1$, $\alpha2\beta1$, $\alpha10\beta1$, and $\alpha11\beta1$ integrins (Barczyk, Carracedo, & Gullberg, 2010; Hynes, 2002; Käpylä et al., 2000; Kern, Eble, Golbik, & Kühn, 1993) but $\alpha1\beta1$ and $\alpha2\beta1$ are the most abundantly expressed (Khoshnoodi et al., 2008; White, Puranen, Johnson, & Heino, 2004).

The two better-studied proteases that degrade Col IV (Page-McCaw et al., 2007) are the gelatinases MMP-2 and MMP-9.

Recently, we showed that surface chemistry influences the pattern of Col IV adsorption, which results in different cellular responses. AFM studies revealed that Col IV self-assembles in specific substratum-dependent patterns, ranging from fine meshworks formed from single molecules (OH and NH_2) to complex networks or molecular aggregates (CH_3 and COOH). Human endothelial cells and fibroblasts bind more avidly to single molecule arrangements of Col IV on hydrophilic (OH) and positively charged (NH_2) surfaces (Coelho et al., 2010, 2011a, 2011b). Finally, we showed that fibroblasts remodeling of adsorbed Col IV depends on material surface properties. Fibroblasts remodeled Col IV by mechanical reorganization, with the involvement of FN, and by enzymatic degradation with involvement of MMP-2 and MMP-9 (Coelho et al., 2013).

In this chapter, we further explore the behavior of Col IV at biomaterials interfaces as it encompasses various concerns in tissue engineering, such as material-induced fibrosis, endothelization, and neovascularization of implants within others (Charonis et al., 2005; Kalluri, 2003; Yurchenco, 2011). To address this, we have employed endothelial cells (HUVECs) as model system considering their involvement in the synthesis, organization, and remodeling of Col IV in the vascular BM (Kalluri & Zeisberg, 2006). We selected the model hydrophobic (CH_3) surface (triggered higher Col IV proteolysis by fibroblasts), and the positively charged NH_2 surface (triggered higher FN fibrillogenesis by fibroblasts). Furthermore, we developed new surfaces combining the two chemical functionalizations at different ratios. We found that HUVEC not only interact with adsorbed Col IV but also remodel it, both mechanically and enzymatically in a way that can be tailored from the materials site, a fact that would favor the tissue engineering application.

2. MATERIAL AND METHODS

2.1 Model Surfaces

2.1.1 Self-Assembled Monolayers

The hydrophobic CH_3 surfaces were prepared as previously described (Coelho et al., 2010; Gustavsson et al., 2008) using an organosilane trichloro(-octadecyl)silane (ODS) purchased from Sigma-Aldrich (Cat. No. 104817). Briefly, the cleaned samples, as above, were placed in a solution containing 12.5 mL of carbon tetrachloride, 37.5 mL of heptane, and 220 μL ODS.

The samples were incubated for 18 min at room temperature and the excess of silane was washed away with pure heptane. Samples were then heated for 1 h at 80°C. For NH_2 functionalization precleaned as above samples were immersed for 18 min at room temperature in a solution containing 30 mL methanol, 10 mL 4% acetic acid glacial, and 3-(2-aminoethylamino) propyltrimethoxysilane ($C_8H_{22}N_2O_3Si$, Sigma-Aldrich, Madrid) to yield a final 1% concentration. Excess of silane was washed by immersion in an excess of solvent solution. Finally, the samples were air dried and heated at 80°C for 1 h.

2.1.2 Mixed NH_2/CH_3 Thiol SAMs

The mixed SAM surfaces were developed as described elsewhere (Keselowsky et al., 2003) using alkanethiols 1-dodecanethiol (HS–$(CH_2)_{11}$–CH_3) (Sigma-Aldrich, Madrid) and 11-amino-1-undecanethiol hydrochloride HS–CH_2–$(CH_2)_9$–CH_2–NH_2 dissolved in HCl (Sigma-Aldrich, Madrid) and deposited on gold (Au). Au-coated glass coverslips (Thermo Fisher, Barcelona) were prepared by deposition of thin films of Ti (150 Å) followed by Au (150 Å) using a high-vacuum evaporator (Polaron E6100) at a deposition rate of 2 Å/s and a chamber base pressure of 2.10^{-6} torr. Glass coverslips were precleaned as earlier at room temperature for 1 h, rinsed consequently with deionized H_2O and 95% ethanol, and dried under a stream of N_2 prior to metal deposition. Freshly prepared Au-coated surfaces were immersed in alkanethiol solutions (1 mM in absolute ethanol) with different ratios (NH_2/CH_3: (0/100, 50/50, 70/30, 90/10, and 100/0)) and allowed to assemble overnight. SAMs were finally rinsed in 95% ethanol and dried under N_2. Surfaces were validated by water contact angle (WCA) measurements using a Dataphysics OCA. The volume of the drop was 20 μL and the measurement was performed after 30 s of substrate–water contact.

2.2 Protein Adsorption

The adsorption of native Col IV (Abcam, UK) on model surfaces was performed at indicated concentrations in 0.1 M sodium acetate (pH 4.5) for 30 min (cellular studies) or 10 min (AFM studies) at 37°C. Adsorption of DQTM collagen type IV (FITC-Col IV, Thermo Fisher) was performed at indicated concentrations in PBS (pH 7.4).

2.2.1 Quantification of Adsorbed FITC-Col IV

The adsorption of FITC-Col IV at different model surfaces was quantified by NaOH extraction as described elsewhere (Coelho et al., 2010, 2011b). Briefly, the model surfaces were cleaned with distilled water in an ultrasonic bath dried and coated for 30 min with 50 μg/mL FITC-Col IV at 37°C. After coating, samples were rinsed three times with PBS. The adsorbed FITC-Col IV was extracted with 250 μL of 0.2 M NaOH for 2 h at room temperature. The fluorescent intensity of the extracts was measured with a fluorescent spectrophotometer (Horiba-Jobin Yvon) set to 488 nm (excitation) and 530 nm (emission), and compared to a standard curve based on known concentrations of FITC-Col IV solutions in 0.2 M NaOH.

2.3 Atomic Force Microscopy

For AFM measurements model surfaces were coated with 10 μg/mL native Col IV as earlier and then dried. We have used the AFM NanoScope III from Digital Instruments (Santa Barbara, CA) operating in the tapping mode in air utilizing Si cantilever from Veeco (Manchester, UK) with force constant of 2.8 N/m and resonance frequency of 75 kHz. The phase signal was set to zero at the resonance frequency of the tip. The tapping frequency was 5–10% lower than the resonance one. Drive amplitude was 200 mV and the amplitude set point Asp was 1.4 V. The ratio between the amplitude set point and the free amplitude was kept equal to 0.7.

2.4 Cells

HUVECs (Labclinics, Barcelona) were culture in supplemented endothelial cell growth medium (Labclinics) containing 0.4% ECGS/H, 2% fetal calf serum, 1 ng/mL epidermal growth factor, 1 mg/mL hydrocortisone, and 1 ng/mL basic fibroblast factor. The medium was exchanged each third day. For the experiments the cells were detached from around confluent flasks with Trypsin/EDTA (Thermo Fisher) and the remained trypsin activity was stopped with 100% fetal bovine serum before two times washing with medium without supplements. Finally, the cells were counted and reconstituted in serum-free endothelial cell of fibroblast medium to be seeded on the different model surfaces.

2.5 Initial Cellular Interaction

2.5.1 Overall Cell Morphology

To study the overall cell morphology we used actin-stained samples. For that purpose, 5×10^4 cells/well were seeded in 24-well tissue culture (TC) plates (Thermo Fisher) containing the samples for 2 h in nonsupplemented endothelial cell medium. Typically, the samples had been precoated with Col IV as earlier and at the end of incubation, the cells were fixed with 4%

paraformaldehyde (20 min), permeabilized with 0.5% Triton X-100 (5 min), and saturated with 1% albumin in PBS (15 min). Actin cytoskeleton was visualized with Alexa Fluor 488 Phalloidin conjugate (Thermo Fisher) and the nuclei with Hoeschst 34580 (Thermo Fisher) dissolved in PBS containing 1% albumin. Finally, the samples were mounted with Mowiol (Sigma-Aldrich), viewed, and photographed on a fluorescent microscope Axio Observer Z1 (Zeiss, Germany) at low magnification (10 × and 20 ×, see later). At least three representative images were acquired for each magnification.

2.5.2 Quantification of Cell Adhesion and Spreading

The number of adhering cells and the mean cell surface area were quantified using the ImageJ plug-ins (National Institute of Health). The adhesion was measured by counting the cells nuclei in at least three randomly chosen squares of each sample (photographed at 10 × magnification). The average cells area was further measured (μm^2) using the same samples but viewed at 20 × magnification in the green channel of the microscope (to visualize cellular actin). The results presented correspond to at least three independent experiments.

2.6 Immunofluorescence

2.6.1 Visualization of Focal Adhesion Contacts

5×10^4 cells/well were seeded as described earlier on Col IV-coated model materials in serum-free medium. The samples were fixed with 4% paraformaldehyde before being permeabilized with 0.5% Triton X-100 (5 min) and saturated in 1% albumin (15 min). Vinculin was visualized using monoclonal anti-vinculin antibody (Sigma-Aldrich) followed by Alexa Fluor 555 goat anti-mouse (Thermo Fisher) as secondary antibody. For most experiments Alexa Fluor 488 Phalloidin conjugate was added to the secondary antibody to visualize actin cytoskeleton together with focal adhesion contacts.

2.6.2 Visualization of Integrins

For the staining of α_1 and α_2 integrins we used monoclonal anti-human integrin α_1 (Millipore, Madrid) or α_2 (Abcam) followed by Alexa Fluor 555 goat anti-mouse as secondary antibody.

2.6.3 Costaining for Vinculin and FAK

To stain vinculin we used the same antibody as earlier while p-FAK was viewed using specific to tyrosine 925 (Tyr 925) polyclonal antibody (Cell Signaling, Barcelona) followed by Alexa Fluor 555 goat anti-rabbit (Thermo

Fisher). Preliminary studies with omitting of the corresponding secondary antibodies were performed to confirm no cross-reactivity in the system.

2.7 Visualization of Adsorbed Collagen IV

To study the fate of adsorbed Col IV on different SAMs, the glass slides (diameter 12 mm) were placed in 24-well TC plates and coated with native Col IV or FITC-Col IV as earlier. After three times washing with PBS, 5×10^4 HUVECs were seeded in nonsupplemented medium and cultured for 5 or 24 h. After 1 h of culture, 10% serum was added to the medium and at the end of incubations all samples were fixed, permeabilized, and saturated as earlier. The samples were immunostained using monoclonal anticollagen IV antibody (Milllipore, Cat. No. MAB1910) followed by Cy3-conjugated goat anti-mouse (Jackson ImmunoResearch, UK) or Alexa Fluor 488 anti-mouse (Thermo Fisher).

2.7.1 Colocalization of Collagen IV with Secreted FN

To study Col IV colocalization with secreted FN cells were cultured on Col IV or FITC-Col IV-coated SAMs for 5 or 24 h as earlier. Native Col IV was visualized as earlier and for both types of Col IV-secreted FN was viewed using polyclonal anti-FN (Sigma-Aldrich) followed by Alexa Fluor 555 anti-rabbit (Thermo Fisher). All samples were viewed and photographed in a fluorescent microscope at higher magnification ($63\times$). At least three representative images were acquired for each experimental condition.

2.8 Western Blot Analysis

2.8.1 FAK Assay

To analyze the expression of p-FAK by western blot 1×10^6 cells were seeded on the model materials (22×40 mm) coated with 50 μg/mL Col IV and further cultured for 2 h in serum-free medium. Cells were lysed in RIPA buffer containing protease and phosphatase inhibitors: 1% Triton X-100, 1% sodium deoxycholate, 0.1% SDS, 150 mM NaCl, 50 mM Tris–HCl (pH 7.5), 1 mM PMSF, 10 mg/mL leupeptin, 10 mg/mL aprotonin, 1% NPO_4, 50 mM NaF, and 200 mM $NaVO_4$. Total protein was quantified using micro-BCA kit (Thermo Fisher). Equal amounts cell lysates were mixed in sample buffer (50 mM Tris–HCl, pH 6.8, 100 mM DTT, 2% SDS, 10% glycerol, and 0.1% bromophenol blue) and separated by SDS-PAGE. After electrotransferring to nitrocellulose membranes (Biorad, Madrid) they were blocked with 5% nonfat dry milk in Tris-saline buffer. The membranes were subsequently incubated in the solution of primary polyclonal antibodies against FAK (dilution 1:1000, Cell Signaling), or p-FAK (Tyr 925) (dilution 1:1000, Cell Signaling) overnight at 4°C. After extensive washing with TBS-Tween (20 mM Tris–HCl, pH 7.6, 137 mM NaCl, 0.1% Tween-20), and pure TBS, HRP-conjugated anti-rabbit secondary antibody (dilution 1:3000, Santa Cruz) was added for 1 h. Protein signal was detected using the ECL chemiluminescent system (GE, Barcelona). Densitometry analysis, standardized to FAK as control for protein loading, was performed using ImageJ.

2.8.2 MMP-2 assay

To detect the expression of MMP-2 in cellular lysates prepared as earlier primary monoclonal antibody against MMP-2 (dilution 1:1000, Santa Cruz) was used. Equal control samples were stained with polyclonal anti-vinculin antibody (dilution 1:1000, Sigma-Aldrich). The corresponding HRP-conjugated secondary antibodies were anti-mouse and anti-rabbit (both from Santa Cruz, Texas), respectively. Densitometry was performed using ImageJ and standardized to the vinculin content as control for protein loading.

2.9 Collagen IV Degradation

2.9.1 Fluorescent Assay

HUVECs were cultured for 5 and 24 h on FITC-Col IV-coated SAMs as earlier and the fluorescence intensity of the culture medium was measured after each culture time. It should be noted that according to the manufacturer's manual this protein (collagen type IV from human placenta origin) is conjugated with FITC in such a condition that part of its fluorescence is quenched; thus, increasing significantly the quantum yield upon extraction under denaturing conditions or when is enzymatically cleaved. Therefore, if cells degrade the adsorbed FITC-Col IV an increase in the fluorescence intensity should be detected in the culture medium. The fluorescent intensity of the extracts and supernatants was measured with a fluorescent spectrophotometer (Horiba-Jobin Yvon, Japan) set to 488 nm (excitation) and 530 nm (emission).

2.9.2 Zymography

The conditioned medium from HUVEC cultured on Col IV-coated model materials for 24 h was used to study the activity of the two major MMPs known to cleave Col IV, namely MMP-2 and MMP-9. For that purpose, 25 μL of each supernatant was mixed with 5 μL of sample buffer (0.04 M

Tris–HCl, pH 6.8, 4% SDS, 33% glycerin, 0.04% bromophenol blue) and these samples were charged on a Ready Gel Zymogram (Biorad, 15-well) containing 10%, gelatin, and subjected to a gel electrophoresis (SDS-PAGE). The gel was then incubated in 2.5% Triton X-100 for 30 min before overnight incubation with renaturation buffer (1 M Tris–HCl, pH 7.5, 5 M NaCl, 1 M $CaCl_2$, 10% Triton X-100) at 37°C. After staining with 0.5% Coomassie brilliant blue R-250 (in 30% methanol/10% acetic acid) and destaining with the same solution without Coomassie, gelatinolytic activity was detected as unstained bands on the blue background of the sample and quantified using a molecular imager gel Doc^+ (imaging system, Biorad).

3. RESULTS AND DISCUSSION

3.1 Remodeling of Col IV by Endothelial Cells on Bare NH_2 and CH_3 Surfaces

As stated in Section 1, vascular BM remodeling involves assembly and degradation of its components and is active during development, angiogenesis, and wound healing. However, it can also occur at the interface of vascular prosthesis with distinct biological consequence. Here, we describe the fate of adsorbed Col IV on model NH_2 and CH_3 surfaces created by SAMs and used alone or mixed in different ratios.

3.2 Morphological Observations

The overall morphology of HUVEC adhering for 2 h on Col IV-coated CH_3 and NH_2 SAMs is shown in Fig. 1. In agreement with our previous investigation (Coelho et al., 2010, 2011b) on CH_3 surface cells demonstrated delayed spreading (Fig. 1A) with short actin filaments, while on NH_2 they were more flattened (Fig. 1B) with prominent actin stress fibers extended in the direction of cell polarization. To learn whether the development of focal adhesion contacts on Col IV induces the recruitment of phosphorylated signaling molecules we costained the samples for vinculin and p-FAK. Fig. 1 shows colocalization between well-developed focal adhesion contacts (green) and p-FAK (red) on NH_2 surfaces (Fig. 1D), resulting in orange when merged. Conversely, on CH_3 surfaces, although some cells were able to make focal contacts (Fig. 1C), a significantly less p-FAK was found, resulting in prevalent green in merges (Fig. 1C). Western blot analysis of cellular lysates is shown in Fig. 1E (upper row). Densitometry showed significantly higher values of p-FAK on NH_2 when compared with

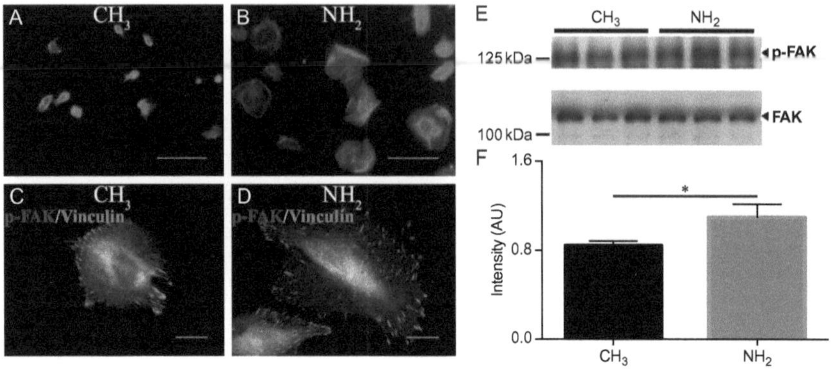

Fig. 1 Initial HUVEC interaction with adsorbed Col IV on CH_3 and NH_2 SAMS. Overall morphology of HUVEC adhering to Col IV-coated CH_3 (A) and NH_2 (B) silane SAMs for 2 h. Bar = 100 μm. Recruitment of p-FAK (C, D—in *red* (*gray* in the print version)) to the focal adhesion complexes (C, D—vinculin in *green* (*light gray* in the print version)) in HUVEC seeded on Col IV-coated CH_3 (C) and NH_2 (D) SAMs. Bar = 20 μm. (E) Western blot of cellular extracts of HUVEC adhering to the same Col IV-coated SAMs for 2 h and correspondent immunoblot for p-FAK and FAK expression. (F) Densitometry analysis of p-FAK expression, standardized to FAK as control for protein loading. *, $P < 0.05$.

CH_3 surface (Fig. 1D). This observation confirms the morphological data, suggesting that HUVEC not only bind more avidly to Col IV on NH_2 surfaces but also transmit better the signal to the cell interior.

3.3 Reorganization and Enzymatic Remodeling of Collagen IV on Bare NH_2 and CH_3 Surfaces

HUVECs were seeded for 5 or 24 h on FITC-Col IV-coated model surfaces. Col IV and secreted FN were viewed simultaneously to confirm the cooperative mechanism of reorganization previously observed for fibroblasts (Coelho et al., 2013). Indeed, as shown in Fig. 2A, the typical fibrillar pattern of secreted FN colocalize almost completely with the substratum arranged FITC-Col IV, resulting in yellow when merged (Fig. 2A e, f) after 5 h of culture. The reorganization, however, is better pronounced on hydrophilic NH_2 substrata (Fig. 2A b, c, and d), while on CH_3 the less Col IV arrangement is combined with the appearance of dark zones of protein removal (Fig. 2A a), corroborating with reduced deposition of FN fibrils (Fig. 2A c). The same trend, but with more pronounced protein removal, was observed after 24 h (Fig. 2A g–h). It is demonstrated by the augmented dark zones on CH_3 substrata where both FN and Col IV are omitted (Fig. 2A g). Increased FN fibrils deposition that colocalized with Col IV in a fibril-like pattern were observed on NH_2 surfaces (Fig. 2A h).

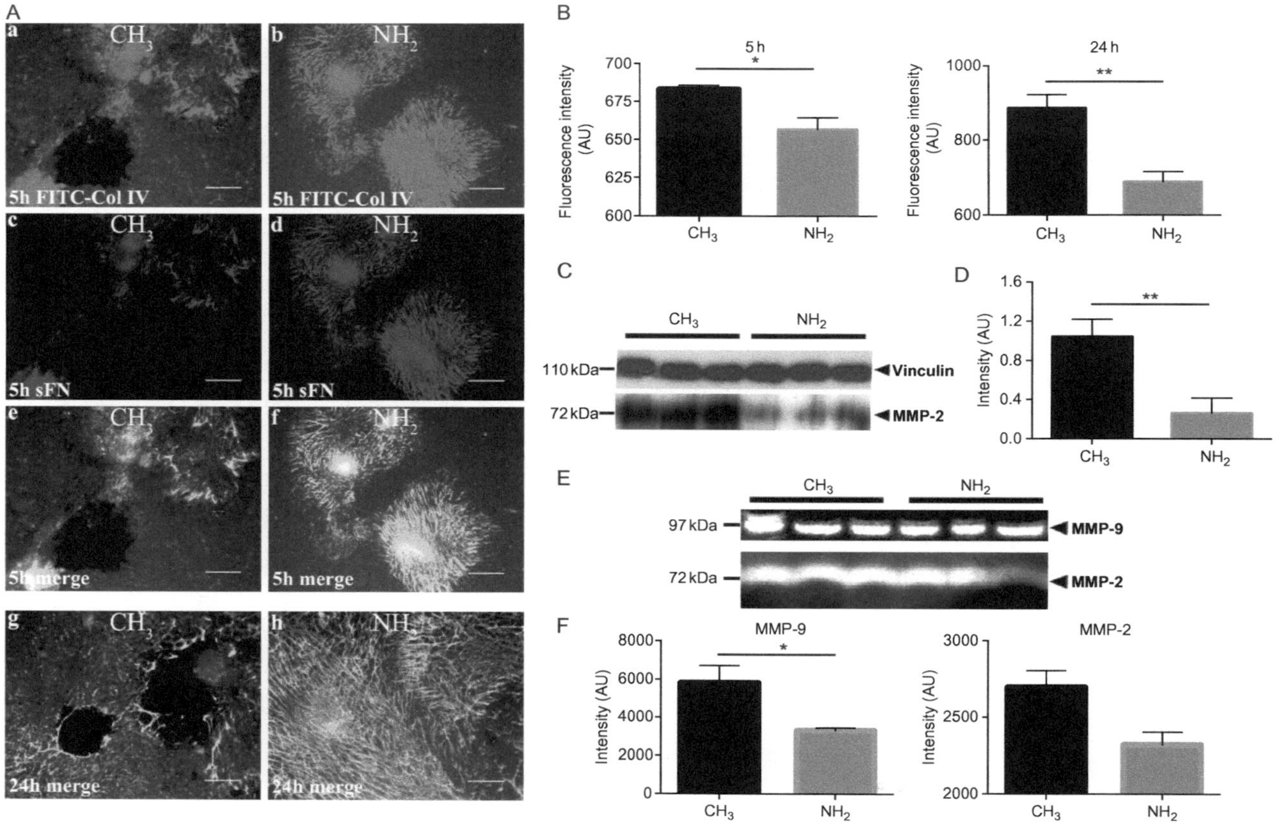

Fig. 2 HUVEC remodeling of adsorbed Col IV on CH_3 and NH_2 SAMs. (A) HUVEC remodeling of adsorbed FITC-Col IV on CH_3 and NH_2 SAMs after 5 (a–f) and 24 h (g, h) corroborated with FN secretion after 5 (c, d—in *red* (*gray* in the print version)) and 24 h. (B) Release of FITC in the supernatant after culturing HUVEC on FITC-Col IV-coated model surfaces for 5 and 24 h. Bar = 20 μm. (C) Western blot of cellular extracts prepared from HUVEC adhering to Col IV-coated model SAMs for 2 h and correspondent immunoblot for MMP-2 and vinculin expression. (D) Densitometry analysis of MMP-2 expression normalized to vinculin as control for protein loading. (E) Zymography of the conditioned medium of HUVEC cultured for 24 h on native Col IV-coated model SAMs. *Line 1*: MMP-9 and *Line 2*: MMP-2, 97 and 72 kDa, respectively. (F) Densitometry for MMP-9 and MMP-2 activity. **, $P < 0.01$; *, $P < 0.05$.

3.4 Quantitative Studies on Enzymatic Remodeling on NH_2 and CH_3 Surfaces

We used FITC-Col IV to quantify the degradation activity of HUVEC toward Col IV assuming that if cells degrade more FITC-Col IV, an increase in the fluorescent intensity in the medium should be detected. Indeed, the quantitative fluorescent measurement of FITC release revealed significantly higher values on CH_3 substrata compared to NH_2 after 5 and 24 h of HUVEC culture. These results support the morphological observations and demonstrate the significantly higher pericellular proteolytic activity of HUVEC toward Col IV on CH_3 substrata (Fig. 2A a).

To further quantify HUVEC proteolytic activity against Col IV, we measured the MMP-2 expression in cell lysates of HUVEC cultured for 24 h on Col IV-coated CH_3 and NH_2 model surfaces. Fig. 2C shows the western blots (left) and the corresponding densitometry for MMP-2 expression normalized to the amount of cellular vinculin. Significantly higher ($P < 0.01$) expression of MMP-2 was found on CH_3 compared to NH_2 surface (Fig. 2C, right). These observations suggest that the type of Col IV assembly on CH_3 substrata, eg, in aggregates (Coelho et al., 2010), triggers significantly higher proteolytic activity of HUVEC.

To follow this observation, we studied the MMP-2 and MMP-9 activity in the conditioned medium after 24 h of HUVEC culture. The zymograms presented in Fig. 2E shows that HUVEC secrete both MMP-2 and MMP-9 supporting their involvement in the remodeling of adsorbed Col IV (Kalluri, 2003). The densitometry analysis (Fig. 2F, bottom) showed higher activity on CH_3 vs NH_2 surfaces ($P < 0.05$ for MMP-9; Fig. 2F, bottom left). Collectively, these data support our morphological observations of higher removal of Col IV by HUVEC on CH_3 surfaces (Fig. 2A).

3.5 Collagen IV Adsorption and Assembly on Mixed NH_2/CH_3 Surfaces

The studies provided earlier supplement our previous investigations (Coelho et al., 2010, 2011b) showing that NH_2 and CH_3 surfaces provide completely different conditions for cellular interaction. Other previous studies also showed that the strongly hydrophobic CH_3 surface suppresses FN matrix organization (Altankov, Grinnell, & Groth, 1996; Altankov & Groth, 1994), while NH_2 which is moderately hydrophilic support FN matrix assembly (Altankov et al., 1996; Altankov & Groth, 1994; Altankov et al., 2010). Considering that FN is involved in Col IV remodeling, we decided to create an order of model surfaces—mixed SAMs—expressing controlled partial ratios of these functionalities. We wanted to explore the possibility of using mixed chemical functionalities to control Col IV remodeling at biomaterials interface.

Mixed SAMs that we used here have been extensively characterized in previous studies (Arima & Iwata, 2007). As a routine control, we have measured the WCA to assess that it is in accordance with published results. WCA decreases as the fraction of amine groups increases from 115 degree on the CH_3 terminated SAM to 20 degree on the NH_2 terminated ones (Fig. 3A).

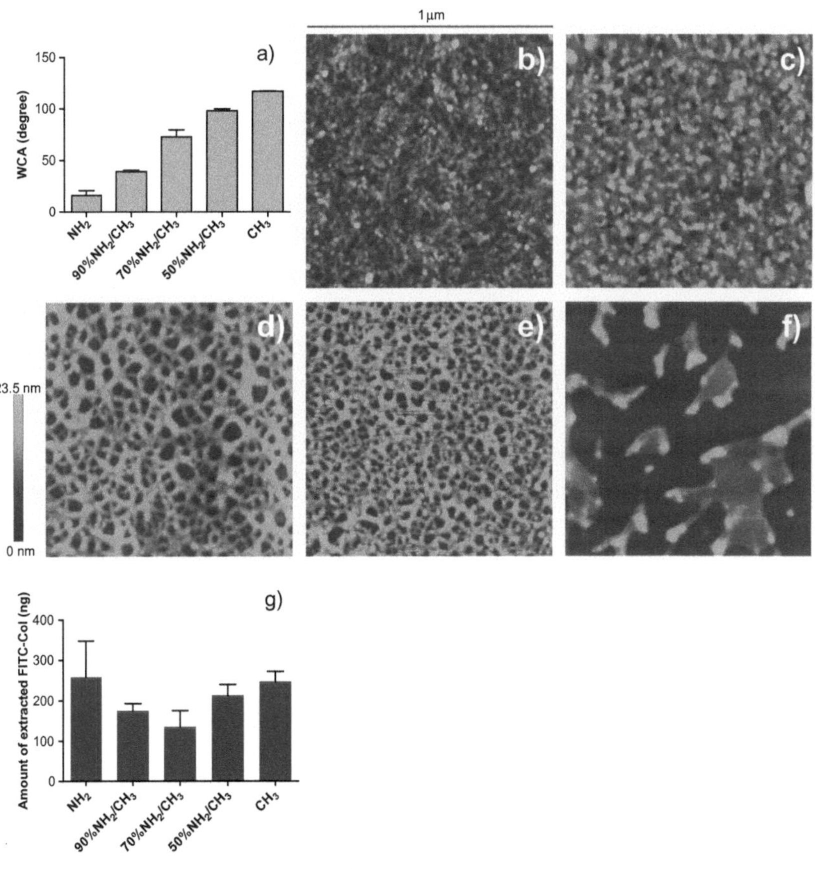

Fig. 3 Collagen IV adsorption and molecular organization on model mixed SAMs. (A) Water contact angle of model thiol SAMs expressing different percentages of NH_2 and CH_3 functionalities. (B–F) AFM images of adsorbed native collagen type IV on model mixed SAMs. Adsorption concentration was 10 mg/mL. (G) Amount of FITC-Col IV extracted from the different mixed SAMs after coating with a 50 mg/mL solution.

The intermediate compositions were justified on the basis that methyl-terminated groups are preferentially incorporated into SAMs over amine-terminated, until its fraction in reaction solutions exceeds 50% (Arima & Iwata, 2007).

Fig. 3B–F shows the AFM images of Col IV adsorbed on the mixed SAMs samples after 10 min from a 10 μg/mL protein solution. The time and concentration of adsorbing solution was chosen based on our previous studies showing that low coating times and low concentrations of absorbing solution allow to better resolve the Col IV surface assembly and distribution as compared to saturation conditions (Coelho et al., 2010, 2011b). As seen in Fig. 3 the NH_2/CH_3 ratio clearly influences the surface assembly and distribution of Col IV. In agreement with our previous work (Coelho et al., 2011b), the surface assembly of Col IV on NH_2 SAMs presented a morphology similar to single-molecule deposition with nonvisible organization but with high-protein deposition. Once the CH_3 groups are incorporated in the mixed SAMs, the Col IV molecules start to form bigger structures (Fig. 3C) that became well-organized and prominent networks, particularly on 70% and 50%NH_2/CH_3 (Fig. 3D and E). Interestingly and also in agreement with our previous findings (Coelho et al., 2010), the Col IV assembly on CH_3 SAM was predominantly in a globular morphology (Fig. 3F) with the appearance of big molecular aggregates, suggesting that the protein–protein interaction overrides the protein–substratum interaction.

FITC-Col IV (50 μg/mL adsorbing solution) was used to study the amount of protein that adsorbs onto the order of mixed SAMs. As shown in Fig. 3G detectable values were obtained for all surfaces, however, no significant differences were found between all mixed SAMs. A clear tendency for higher adsorption on pure NH_2 and CH_3 surfaces was observable suggesting that the mixed functionalities disturb the driving forces that control Col IV adsorption onto surfaces of single chemistry.

3.6 HUVEC Interaction with Collagen IV on Mixed NH_2/CH_3 Surfaces

Fig. 4A shows the overall morphology of HUVEC adhering for 2 h on plane (a–e) and Col IV coated (f–j) model mixed SAMs. On plane mixed SAMs, significantly less cells adhered compared to Col IV coated mixed SAMs (Fig. 4A and B). Accordingly, all cells cultured on Col IV coated mixed SAMs showed significantly higher cell surface area when compared to respective bare surface (Fig. 4A and B). Furthermore, HUVEC cultured

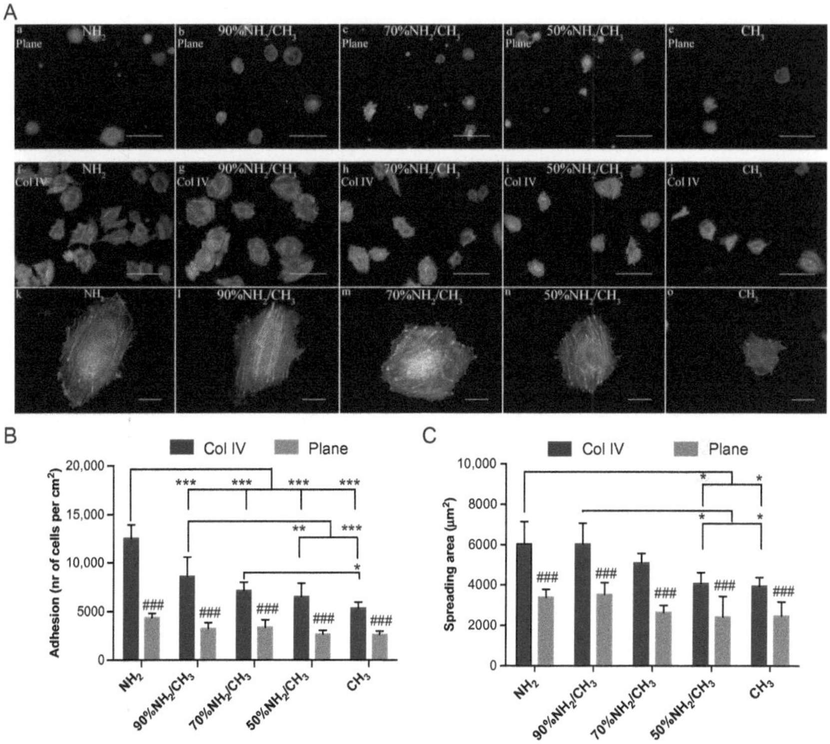

Fig. 4 Initial interaction of HUVEC with plane and Col IV coated mixed SAMs. (A) Overall morphology of HUVEC adhering to plane (a–e) and Col IV coated (f–j) mixed SAMs. The cells were stained for actin and nucleus after 2 h of culture. Bar = 100 μm. Development of focal adhesion contacts (k–o vinculin in *red* (*gray* in the print version)) and actin cytoskeleton viewed by FITC phalloidin in *green* (*light gray* in the print version) of HUVEC seeded on Col IV coated mixed SAMs for 2 h. Only merged images are shown. Bar = 20 μm. (B) Represents the quantities for cell adhesion expressed as number of cells per cm^2 and (C) spreading in μm^2 for both plane (*gray bars*) and Col IV-coated (*black bars*) model mixed SAMs. $^{\#}$, significantly different from Col IV-coated SAMs; *, significantly different between Col IV-coated SAMs. *** and $^{\#\#\#}$, $P < 0.0001$; **, $P < 0.01$; *, $P < 0.05$.

on Col IV-coated NH_2, and 90%NH_2/CH_3, presented significantly bigger cell surface areas when compared to Col IV coated 50%NH_2/CH_3 and pure CH_3 SAMs (Fig 4A and C). The more flattened and elongated cells on all NH_2-containing SAMs presented well-developed focal adhesion contacts (viewed by vinculin in red) where prominent actin stress fibers (green) often inserts, as seen on merged high magnification images on the bottom row (Fig. 4A, k–p). To learn which integrins and at what extent were involved in the adhesion of HUVEC we studied the expression and organization of

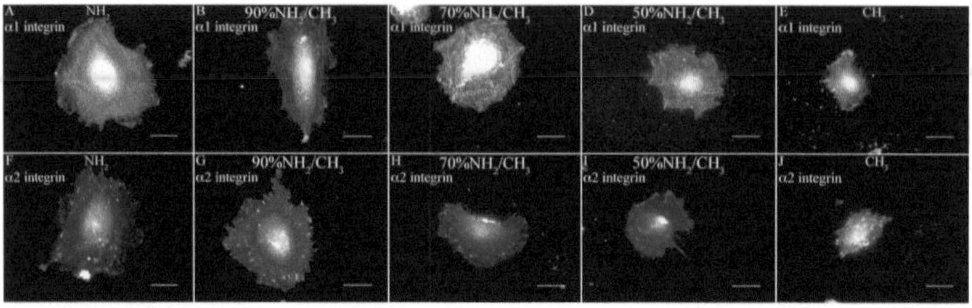

Fig. 5 HUVEC signaling on Col IV-coated model mixed SAMs. Expression of alpha 1 (A–E) and alpha 2 (F–J) integrins in HUVEC adhering on the same mixed SAMs coated with Col IV. Bar 20 μm.

both α_1 and α_2 integrins. As shown in Fig. 5 both α_1 and α_2 integrins are expressed by HUVEC cultured of Col IV coated mixed SAMs (A–E and F–J). However, the expression diminishes as the relative amount of CH_3 groups increase (A–E and F–J) becoming lowest on pure CH_3 (E, J). Alpha 1 represents a rather diffuse pattern well seen on most SAMs (A–C), while α_2 localize in structures resembling focal adhesion contacts (F–J), which also diminish in size with increasing the ratio of CH_3 groups (F–J).

3.7 Collagen IV Remodeling on Mixed NH_2/CH_3 SAMs

HUVECs were seeded for 5 or 24 h on FITC-Col IV coated mixed SAMs and both Col IV reorganization and FN secretion were viewed by direct or indirect immunofluorescence. Fig. 6 shows the typical organization of FN fibrils at fifth hour (Fig. 6A f–g) which colocalize almost completely with linearly arranged Col IV (Fig. 6A a–f) seen in orange on merged images (Fig. 6A k–o). The reorganization of both proteins generally diminishes when the relative amount of CH_3 groups increases. Therefore, the worst Col IV arrangement was observed on pure CH_3 surface (Fig. 6A e) which corroborates with the lower deposition of FN fibrils (Fig. 6A j) and with the increase of dark zones of protein removal (Fig. 6A e). The same trend in Col IV remodeling along with FN matrix formation was observed after 24 h, as shown in Fig. 6A (p–t) presenting only merged images.

3.8 Quantitative Studies on Enzymatic Collagen IV Remodeling

Studies on FITC-Col IV release in the supernatant revealed little differences after 5 h of HUVEC culture; nevertheless, significantly higher values were found on CH_3 compared to $90\%NH_2/CH_3$ surfaces (Fig. 6B). After 24 h, the signal from CH_3 sample was already significantly higher vs all other surfaces. Fig. 7 shows HUVEC remodeling of native Col IV after 24 h viewed

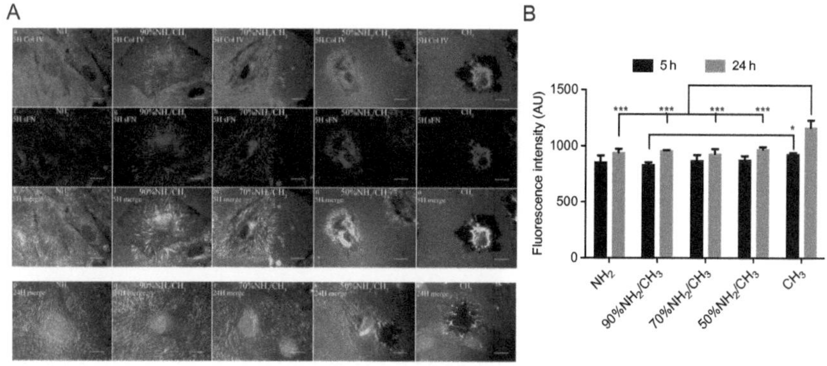

Fig. 6 Remodeling of native Col IV by HUVEC on mixed SAMs. (A) Remodeling of adsorbed Col IV by HUVEC on different mixed SAMs for 5 h (a–e) and 24 h (p–t) viewed simultaneously with secreted FN at 5 h (f–j) and 24 h where only merged images are shown (k–o, p–t). Bar = 20 μm. (B) FITC signal detected in the supernatants of HUVEC cultured for 5 and 24 h on FITC-Col IV coated mixed SAMs. The released fluorescent signal is correlated with the proteolytic activity of cells. ***, $P < 0.0001$; *, $P < 0.05$.

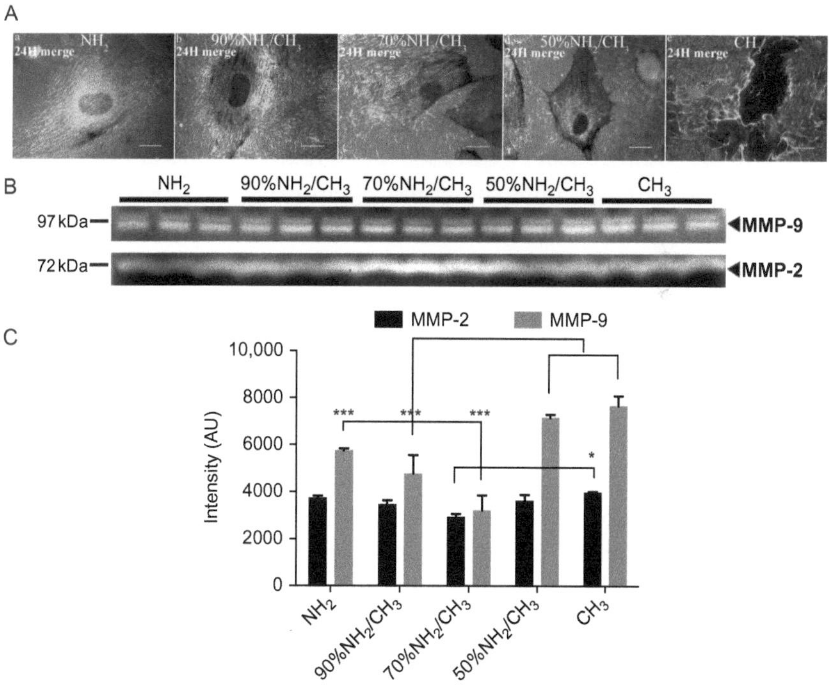

Fig. 7 (A) Remodeling of adsorbed native Col IV (a–e—in *green* (*light gray* in the print version)) by HUVEC after 24 h viewed simultaneously with secreted FN (a–e—in *red* (*gray* in the print version)). Only merged images are shown. Bar = 20 μm. (B) Zymography of the conditioned medium of HUVEC cultured for 24 h on Col IV coated mixed SAMs. *Upper line* represents MMP-9 and the *lower* MMP-2, 97 and 72 kDa, respectively. (C) Shows the densitometry values for MMP-9 and for MMP-2 activity. ***, $P < 0.0001$.

by immunofluorescence together with secreted FN. Visibly, there were no differences in reorganization compared to FITC-Col IV-coated surfaces (Fig. 6). Again the general trend was the increased FN matrix formation as the percentage of NH_2 groups increase, and higher degradation of adsorbed Col IV with increasing the percentage of CH_3 groups.

The conditioned medium of HUVEC adhering on Col IV on the different samples was collected after 24 h and used for zymography analysis to quantify the activity of MMP-2 and MMP-9. Zymograms presented in Fig. 7B show that HUVECs secrete both MMPs in all conditions, but only MMP-9 seems to be secreted in both inactive and active forms (two bands). MMP-2 was found only in inactive form (one band). The densitometry for MMP-9 (Fig. 7C) showed significantly higher levels for CH_3 and 50%NH_2/CH_3 samples when compared with the other surfaces. The densitometry for MMP-2 (Fig. 7C) showed significant differences only between 70%NH_2/CH_3 and pure CH_3 surfaces.

4. CONCLUSIVE REMARKS

This study demonstrates that human endothelial cells not only interact well with the adsorbed Col IV but also tend to remodel it in a distinct pattern depending on the underlining substratum chemistry. Two types of cell activities might be foreseen: Col IV reorganization and Col IV enzymatic degradation. In agreement with previous investigation (Coelho et al., 2013; Maneva-Radicheva et al., 2008) we show that cells are able to mechanically translocate adsorbed Col IV in fibril-like pattern that frequently colocalize with FN matrix fibrils shown to be better formed on hydrophilic substrata (Altankov & Groth, 1996; Grinnell, 1986). An important question is whether this process of Col IV reorganization is physiologically relevant, as it is observed in 2D conditions where endothelial cells behave differently from their natural 3D environment during angiogenesis. Moreover, it is noteworthy that Col IV is a nonfibrillar matrix protein typically observed in the BM, where it assembles in a sheet-like structure providing the major structural support for epithelial cells (Kalluri, 2003; Khoshnoodi et al., 2008; Kühn, 1995; Yurchenco, 2011). The way of its spatial arrangement in the BM is not clear but it is widely accepted that starts by self-assembly through NC1 domain forming dimmers and through 7S forming tetramers. Then extensive disulfide bounding gives rise to the superstructure found in the BMs (Kalluri, 2003; Khoshnoodi et al., 2008). From this angle the rearrangement of adsorbed Col IV in 2D substrata does not look physiologically relevant, though it obviously utilize a cell-dependent mechanism. It should be noticed, however, that similar linear organization of Col IV has been observed during early BM assembly in an in vitro 3D skin culture model (Fleischmajer et al., 1998), shown to be also cell-driven process. In addition, our previous investigations on Col IV remodeling in contact with cancer cells also demonstrate the involvement of $\alpha 2\beta 1$ integrins in the process (Maneva-Radicheva et al., 2008).

Notwithstanding concerns on its physiological relevance, our data suggest that the driving force for Col IV reorganization is its association with FN fibrils. In fact, such mechanism is demonstrated for fibrillar collagens type I and III (Kadler, Hill, & Canty-Laird, 2008; Velling, Risteli, Wennerberg, Mosher, & Johansson, 2002), and also for adsorbed fibrinogen (Tzoneva et al., 2002), but we are the first showing that it works also for Col IV. Thus, our study provides a causal link between the protein conformation and the endothelial cells interaction. Collectively, it suggests that distinct chemical functionalization, like for example NH_2, combined with Col IV pre-adsorption comprises a prospective tool for the engineering of biomimetic interfaces that might improve the endothelization of cardiovascular implants.

REFERENCES

Abrahamsson, I., Zitzmann, N. U., Berglundh, T., Linder, E., Wennerberg, A., & Lindhe, J. (2002). The mucosal attachment to titanium implants with different surface characteristics: An experimental study in dogs. *Journal of Clinical Periodontology, 29*, 448–455.

Ala-aho, R., & Kähäri, V.-M. (2005). Collagenases in cancer. *Biochimie, 87*, 273–286.

Altankov, G., Grinnell, F., & Groth, T. (1996). Studies on the biocompatibility of materials: Fibroblast reorganization of substratum-bound fibronectin on surfaces varying in wettability. *Journal of Biomedical Materials Research, 30*, 385–391.

Altankov, G., & Groth, T. (1994). Reorganization of substratum-bound fibronectin on hydrophilic and hydrophobic materials is related to biocompatibility. *Journal of Materials Science: Materials in Medicine, 5*, 732–737.

Altankov, G., & Groth, T. (1996). Fibronectin matrix formation and the biocompatibility of materials. *Journal of Materials Science: Materials in Medicine*, 7, 425–429.

Altankov, G., Groth, T., Engel, E., Gustavsson, J., Pegueroles, M., Aparicio, C., et al. (2010). Development of provisional extracellular matrix on biomaterials interface: Lessons from in vitro cell culture. In V. P. Shastri, G. Altankov, & A. Lendlein (Eds.), *Advances in Regenerative Medicine: Role of Nanotechnology, and Engineering Principles* (pp. 19–43). Netherlands: Springer.

Arima, Y., & Iwata, H. (2007). Effect of wettability and surface functional groups on protein adsorption and cell adhesion using well-defined mixed self-assembled monolayers. *Biomaterials, 28*, 3074–3082.

Aumailley, M., & Gayraud, B. (1998). Structure and biological activity of the extracellular matrix. *Journal of Molecular Medicine (Berlin), 76*, 253–265.

Barczyk, M., Carracedo, S., & Gullberg, D. (2010). Integrins. *Cell and Tissue Research, 339*, 269–280.

Charonis, A., Sideraki, V., Kaltezioti, V., Alberti, A., Vlahakos, D., Wu, K., et al. (2005). Basement membrane peptides: Functional considerations and biomedical applications in autoimmunity. *Current Medicinal Chemistry, 12*, 1495–1502.

Coelho, N. M., Gonzalez-Garcia, C., Planell, J. A., Salmeron-Sanchez, M., & Altankov, G. (2010). Different assembly of type IV collagen on hydrophilic and hydrophobic substrata alters endothelial cells interaction. *European Cells & Materials*, *19*, 262–272.

Coelho, N. M., Gonzalez-Garcia, C., Salmeron-Sanchez, M., & Altankov, G. (2011a). Arrangement of type IV collagen and laminin on substrates with controlled density of –OH groups. *Tissue Engineering Part A*, *17*, 2245–2257.

Coelho, N. M., Gonzalez-Garcia, C., Salmeron-Sanchez, M., & Altankov, G. (2011b). Arrangement of type IV collagen on NH2 and COOH functionalized surfaces. *Biotechnology and Bioengineering*, *108*, 3009–3018.

Coelho, N. M., Salmeron-Sanchez, M., & Altankov, G. (2013). Fibroblasts remodeling of type IV collagen at a biomaterials interface. *Biomaterials Science*, *1*, 494–502.

Cox, T. R., & Erler, J. T. (2011). Remodeling and homeostasis of the extracellular matrix: Implications for fibrotic diseases and cancer. *Disease Models & Mechanisms*, *4*, 165–178.

Daley, W. P., Peters, S. B., & Larsen, M. (2008). Extracellular matrix dynamics in development and regenerative medicine. *Journal of Cell Science*, *121*, 255–264.

Fleischmajer, R., Perlish, J. S., Macdonald, D. E., Schechter, A., Murdoch, A. D., Iozzo, R. V., et al. (1998). There is binding of collagen IV to β1 integrin during early skin basement membrane assembly. *Annals of the New York Academy of Sciences*, *857*, 212–227.

Frantz, C., Stewart, K. M., & Weaver, V. M. (2010). The extracellular matrix at a glance. *Journal of Cell Science*, *123*, 4195–4200.

García, A. J. (2005). Get a grip: Integrins in cell–biomaterial interactions. *Biomaterials*, *26*, 7525–7529.

Geiger, B., Bershadsky, A., Pankov, R., & Yamada, K. M. (2001). Transmembrane crosstalk between the extracellular matrix and the cytoskeleton. *Nature Reviews Molecular Cell Biology*, *2*, 793–805.

Geiger, B., Spatz, J. P., & Bershadsky, A. D. (2009). Environmental sensing through focal adhesions. *Nature Reviews. Molecular Cell Biology*, *10*, 21–33.

Goody, M. F., & Henry, C. A. (2010). Dynamic interactions between cells and their extracellular matrix mediate embryonic development. *Molecular Reproduction and Development*, *77*, 475–488.

Grinnell, F. (1986). Focal adhesion sites and the removal of substratum-bound fibronectin. *The Journal of Cell Biology*, *103*, 2697–2706.

Grinnell, F., & Feld, M. K. (1982). Fibronectin adsorption on hydrophilic and hydrophobic surfaces detected by antibody binding and analyzed during cell adhesion in serum-containing medium. *Journal of Biological Chemistry*, *257*, 4888–4893.

Gustavsson, J., Altankov, G., Errachid, A., Samitier, J., Planell, J., & Engel, E. (2008). Surface modifications of silicon nitride for cellular biosensor applications. *Journal of Materials Science. Materials in Medicine*, *19*, 1839–1850.

Hubbell, J. A. (2003). Materials as morphogenetic guides in tissue engineering. *Current Opinion in Biotechnology*, *14*, 551–558.

Hynes, R. O. (2002). Integrins: Bidirectional, allosteric signaling machines. *Cell*, *110*, 673–687.

Hynes, R. O. (2009). The extracellular matrix: Not just pretty fibrils. *Science*, *326*, 1216–1219.

Irvine, S. M., Cayzer, J., Todd, E. M., Lun, S., Floden, E. W., Negron, L., et al. (2011). Quantification of in vitro and in vivo angiogenesis stimulated by ovine forestomach matrix biomaterial. *Biomaterials*, *32*, 6351–6361.

Janmey, P. A., & McCulloch, C. A. (2007). Cell mechanics: Integrating cell responses to mechanical stimuli. *Annual Review of Biomedical Engineering*, *9*, 1–34.

Kadler, K. E., Hill, A., & Canty-Laird, E. G. (2008). Collagen fibrillogenesis: Fibronectin, integrins, and minor collagens as organizers and nucleators. *Current Opinion in Cell Biology, 20*, 495–501.

Kalluri, R. (2003). Basement membranes: Structure, assembly and role in tumour angiogenesis. *Nature Reviews. Cancer, 3*, 422–433.

Kalluri, R., & Zeisberg, M. (2006). Fibroblasts in cancer. *Nature Reviews Cancer, 6*, 392–401.

Käpylä, J., Ivaska, J., Riikonen, R., Nykvist, P., Pentikäinen, O., Johnson, M., et al. (2000). Integrin α2I domain recognizes Type I and Type IV collagens by different mechanisms. *Journal of Biological Chemistry, 275*, 3348–3354.

Keresztes, Z., Rouxhet, P. G., Remacle, C., & Dupont-Gillain, C. (2006). Supramolecular assemblies of adsorbed collagen affect the adhesion of endothelial cells. *Journal of Biomedical Materials Research. Part A, 76A*, 223–233.

Kern, A., Eble, J., Golbik, R., & Kühn, K. (1993). Interaction of type IV collagen with the isolated integrins α1β1 and α2β1. *European Journal of Biochemistry, 215*, 151–159.

Keselowsky, B. G., Collard, D. M., & García, A. J. (2003). Surface chemistry modulates fibronectin conformation and directs integrin binding and specificity to control cell adhesion. *Journal of Biomedical Materials Research. Part A, 66A*, 247–259.

Keselowsky, B. G., Collard, D. M., & García, A. J. (2004). Surface chemistry modulates focal adhesion composition and signaling through changes in integrin binding. *Biomaterials, 25*, 5947–5954.

Khoshnoodi, J., Pedchenko, V., & Hudson, B. G. (2008). Mammalian collagen IV. *Microscopy Research and Technique, 71*, 357–370.

Kolahi, K. S., & Mofrad, M. R. (2010). Mechanotransduction: A major regulator of homeostasis and development. *Wiley Interdisciplinary Reviews. Systems Biology and Medicine, 2*, 625–639.

Kühn, K. (1995). Basement membrane (type IV) collagen. *Matrix Biology, 14*, 439–445.

Larsen, M., Artym, V. V., Green, J. A., & Yamada, K. M. (2006). The matrix reorganized: Extracellular matrix remodeling and integrin signaling. *Current Opinion in Cell Biology, 18*, 463–471.

Llopis-Hernández, V., Rico, P., Ballester-Beltrán, J., Moratal, D., & Salmerón-Sánchez, M. (2011). Role of surface chemistry in protein remodeling at the cell-material interface. *PLoS One, 6*, e19610.

Lu, P., Takai, K., Weaver, V. M., & Werb, Z. (2011). Extracellular matrix degradation and remodeling in development and disease. *Cold Spring Harbor Perspectives in Biology, 3*, 1–24.

Maneva-Radicheva, L., Ebert, U., Dimoudis, N., & Altankov, G. (2008). Fibroblast remodeling of adsorbed collagen type IV is altered in contact with cancer cells. *Histology and Histopathology, 23*, 833–842.

Moon, I. S., Berglundh, T., Abrahamsson, I., Linder, E., & Lindhe, J. (1999). The barrier between the keratinized mucosa and the dental implant. *Journal of Clinical Periodontology, 26*, 658–663.

Page-McCaw, A., Ewald, A. J., & Werb, Z. (2007). Matrix metalloproteinases and the regulation of tissue remodelling. *Nature Reviews. Molecular Cell Biology, 8*, 221–233.

Place, E. S., Evans, N. D., & Stevens, M. M. (2009). Complexity in biomaterials for tissue engineering. *Nature Materials, 8*, 457–470.

Ratner, B. D., & Bryant, S. J. (2004). Biomaterials: Where we have been and where we are going. *Annual Review of Biomedical Engineering, 6*, 41–75.

Shi, F., Harman, J., Fujiwara, K., & Sottile, J. (2010). Collagen I matrix turnover is regulated by fibronectin polymerization. *American Journal of Physiology. Cell Physiology, 298*, C1265–C1275.

Stamenkovic, I. (2003). Extracellular matrix remodelling: The role of matrix metalloproteinases. *The Journal of Pathology, 200*, 448–464.

Sternlicht, M. D., & Werb, Z. (2001). How matrix metalloproteinases regulate cell behavior. *Annual Review of Cell and Developmental Biology*, *17*, 463–516.

Stevens, M. M., & George, J. H. (2005). Exploring and engineering the cell surface interface. *Science*, *310*, 1135–1138.

Thevenot, P., Hu, W., & Tang, L. (2008). Surface chemistry influences implant biocompatibility. *Current Topics in Medicinal Chemistry*, *8*, 270–280.

Timpl, R., & Brown, J. C. (1996). Supramolecular assembly of basement membranes. *Bioessays*, *18*, 123–132.

Toromanov, G., Gugutkov, D., Gustavsson, J., Planell, J., Salmerón-Sánchez, M., & Altankov, G. (2015). Dynamic behavior of vitronectin at the cell–material interface. *ACS Biomaterials Science & Engineering*, *1*, 927–934.

Tzoneva, R., Groth, T., Altankov, G., & Paul, D. (2002). Remodeling of fibrinogen by endothelial cells in dependence on fibronectin matrix assembly. Effect of substratum wettability. *Journal of Materials Science. Materials in Medicine*, *13*, 1235–1244.

Velling, T., Risteli, J., Wennerberg, K., Mosher, D. F., & Johansson, S. (2002). Polymerization of type I and III collagens is dependent on fibronectin and enhanced by integrins $\alpha 11\beta 1$ and $\alpha 2\beta 1$. *Journal of Biological Chemistry*, *277*, 37377–37381.

White, D. J., Puranen, S., Johnson, M. S., & Heino, J. (2004). The collagen receptor subfamily of the integrins. *The International Journal of Biochemistry & Cell Biology*, *36*, 1405–1410.

Wynn, T. A. (2008). Cellular and molecular mechanisms of fibrosis. *The Journal of Pathology*, *214*, 199–210.

Yamada, K. M., & Geiger, B. (1997). Molecular interactions in cell adhesion complexes. *Current Opinion in Cell Biology*, *9*, 76–85.

Yurchenco, P. D. (2011). Basement membranes: Cell scaffoldings and signaling platforms. *Cold Spring Harbor Perspectives in Biology*, *3*.

Yurchenco, P., & Schittny, J. (1990). Molecular architecture of basement membranes. *The FASEB Journal*, *4*, 1577–1590.

CHAPTER FIVE

Cytochromes P450: History, Classes, Catalytic Mechanism, and Industrial Application

D.J. Cook*,†, J.D. Finnigan*,†, K. Cook‡, G.W. Black†,[1], S.J. Charnock*,[1]
*Prozomix Limited, Haltwhistle, Northumberland, United Kingdom
†Faculty of Health and Life Sciences, Northumbria University, Newcastle upon Tyne, United Kingdom
‡Thermo Fisher Scientific, Hemel Hempstead, Hertfordshire, United Kingdom
[1]Corresponding authors: e-mail address: gary.black@northumbria.ac.uk; simon.charnock@prozomix.com

Contents

1. Background and History of Cytochromes P450 106
2. P450 Classes 110
3. P450 Family Nomenclature 114
4. Catalytic Cycle 115
5. Uncoupling 117
6. P450s in Industry 119
References 123

Abstract

Cytochromes P450, a family of heme-containing monooxygenases that catalyze a diverse range of oxidative reactions, are so-called due to their maximum absorbance at 450 nm, ie, "Pigment–450 nm," when bound to carbon monoxide. They have appeal both academically and commercially due to their high degree of regio- and stereoselectivity, for example, in the area of active pharmaceutical ingredient synthesis. Despite this potential, they often exhibit poor stability, low turnover numbers and typically require electron transport protein(s) for catalysis. P450 systems exist in a variety of functional domain architectures, organized into 10 classes. P450s are also divided into families, each of which is based solely on amino acid sequence homology. Their catalytic mechanism employs a very complex, multistep catalytic cycle involving a range of transient intermediates. Mutagenesis is a powerful tool for the development of improved biocatalysts and has been used extensively with the archetypal Class VIII P450, BM3, from *Bacillus megaterium*, but with the increasing scale of genomic sequencing, a huge resource is now available for the discovery of novel P450s.

Advances in Protein Chemistry and Structural Biology, Volume 105
ISSN 1876-1623
http://dx.doi.org/10.1016/bs.apcsb.2016.07.003

1. BACKGROUND AND HISTORY OF CYTOCHROMES P450

Cytochromes P450 (P450s) are a family of heme-containing monooxygenase enzymes which perform a diverse range of oxidative reactions against a highly diverse range of substrates. They activate molecular oxygen, incorporating one atom into an organic substrate, while simultaneously using the other for the production of a water molecule (Hannemann, Bichet, Ewen, & Bernhardt, 2007). This mechanism is utilized in an array of catalytic reactions, most notably that of hydroxylations. Others, however, include epoxidations, deaminations, and dealkylations (Bernhardt, 2006; Fujii et al., 2006; Podust & Sherman, 2012). To add to this, P450s demonstrate a high degree of regio- and stereoselectivity, a trait which produces great interest within the field of pharmaceutical chemistry (Niwa, Murayama, & Yamazaki, 2011).

The manufacture of active pharmaceutical ingredients (APIs) is a task in which it is considered that P450s could play a very important role (Niwa et al., 2011). Current chemical methods of producing such intermediates are generally very wasteful, using an excess of reagents, temperature, and pressure which results in significant metal and solvent waste. P450s, in common with other biocatalysis enzymes, do not require this, giving them real green credentials and producing increasing interest toward the elucidation of their true capabilities (Urlacher, 2010).

The first experimental evidence for cytochromes P450 came in the mid-1950s (Brodie et al., 1955) where an enzyme system within liver microsomes of rabbits was found to oxidize xenobiotic compounds. However, it was not until 1958 that Klingenberg formally identified and named P450s, when it was fortuitously discovered that bubbling carbon monoxide through a preparation of rat liver microsomes resulted in a spectrum with maximum absorbance at 450 nm (Klingenberg, 1958). It is this maximum absorbance which gave P450s their name, ie, "Pigment–450 nm."

Much has since been learnt; in 1964, Omura and Sato deduced the P450s to be type-*b* hemoproteins, meaning that they contain a protoporphyrin IX ring structure. A porphyrin ring is a hydrophobic, planar molecule which contains four pyrrole subunits connected by methine bridges (Fig. 1).

This structure produces four inward facing nitrogen atoms which coordinate a central, reduced iron atom (Fe^{2+}). The iron atom has five possible coordination bonds, four of which are bound to the surrounding nitrogen

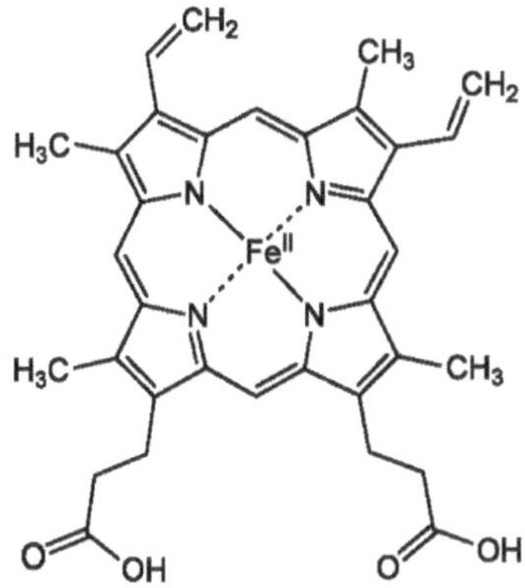

Fig. 1 Type-*b* heme. Heme group showing structure of protoporphyrin ring with centrally bound iron atom.

atoms and one of which is bound to an axial thiol ligand. Omura and Sato (1964) used the isolate of rabbit liver microsomes, where they reduced the P450 in solution using sodium dithionite and then observed the difference spectra when the protein was bound to either carbon monoxide or ethyl-isocyanide, in both aerobic and anaerobic conditions. They compared their observations closely to cytochrome b_5, a known hemoprotein. Their results proved P450s to be heme containing, while noting key differences in spectra and thus concluding P450s to be a novel enzyme class utilizing a novel heme prosthetic group.

In 1968, the Gunsalus group discovered cytochrome P450cam in the Gram-negative saprotrophic soil bacterium *Pseudomonas putida* and named it so for its activity, the 5-*exo*-hydroxylation of camphor (Katagiri, Ganguli, & Gunsalus, 1968), which can be used by this organism as a sole carbon source (Grogan, Roberts, Wan, & Willetts, 1993). Later, in 1977, the same group used resonance Raman spectroscopy to confirm that the axial thiol residue was in fact a cysteine residue, helping to explain much of the unusual Soret absorption peaks indicative of P450 enzymes previously observed by Omura and Sato (1964).

In 1985, the Gunsalus group continued their work with P450cam and in collaboration with two other groups produced the first crystal structure from the P450 superfamily (Fig. 2; Poulos et al., 1985). The data produced during this pivotal study confirmed many of the previous observations and assumptions regarding this enzyme family; it identified that approximately 40% of the structure is made up of 12 helical segments, designated A–L, and that unlike many other enzymes solved at that time, it does not fold into discrete

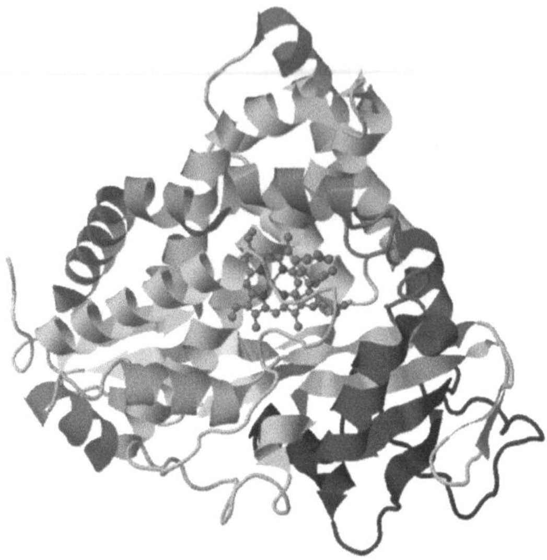

Fig. 2 Crystal structure of P450cam. Crystal structure of P450cam (PDB: 2CPP) solved at a resolution of 1.63 Å (Poulos, Finzel, GunsalusII, Wagner, & Kraut, 1985). The heme group can be seen located in the center.

N- and C-terminal domains and lacks any cleft or hinge. It instead comprises helix-rich and helix-poor domains. The protoporphyrin ring is embedded between two of these helices (I and L), and due to the structure being solved in the presence of camphor, a number of critical amino acid residues were identified. The most noteworthy is Cys^{357}, which was shown to be responsible for the formation of the thiol ligand, extending in from the N-terminal end of the proximal helix (I). This feature is at odds with a number of other heme-containing proteins such as hemoglobin, myoglobin, and cytochrome *c*, all of which employ a histidine residue which extends in from the C-terminal side of the proximal helix (Poulos et al., 1985). It is now known that this cysteine is a fully conserved residue in almost all P450 enzymes (Oezguen et al., 2008; Raucy & Allen, 2001; Rupasinghe et al., 2006).

A number of residues were shown to be important for the binding of camphor within the active site pocket (Poulos et al., 1985); Tyr^{96} was identified as being responsible for anchoring the substrate via a hydrogen bond, formed between the hydroxyl group of this residue and the carbonyl oxygen atom of camphor. A number of other residues were shown to come into play, orientating the substrate using a series of hydrophobic interactions, namely Phe^{87}, Leu^{244}, Val^{247}, Thr^{252}, and Val^{295} (Fig. 3). The location of the heme prosthetic group also shed light on the necessity of P450cam to

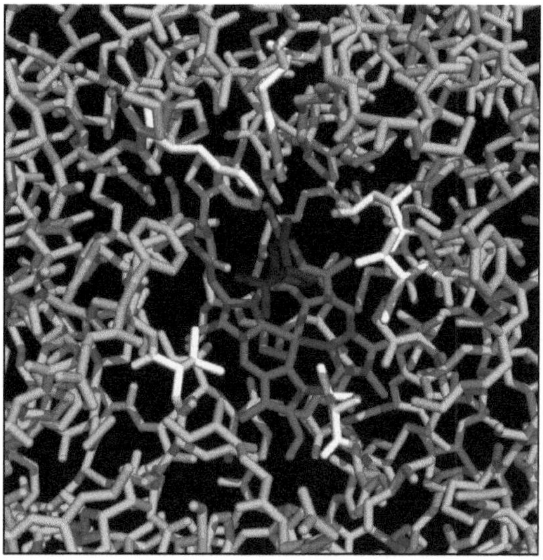

Fig. 3 Crystal structure of P450cam active site. Crystal structure (1.63 Å) of the P450cam active site bound with camphor (*blue* (*dark gray* in the print version)), with important residues highlighted (*green* (*light gray* in the print version); Poulos et al., 1985; PDB: 2CPP). The fully conserved cysteine can also be seen (*yellow* (*gray* in the print version)). A number of residues were removed to aid in visualization.

employ electron transport proteins to provide the reducing equivalents needed for catalysis, as it was shown to be buried within the helix-rich domain and therefore unable to directly accept electrons from NADH, its obligate donor (Poulos et al., 1985).

Since this early work on P450cam, P450s have now been demonstrated to appear in all orders of life, with almost all organisms having at least one example. There are some notable exceptions, however, such as *Escherichia coli* that lacks any identifiable P450 homologue in its entire genome, an observation also true for some other microbes (Guengerich & Cheng, 2011).

The advent and commercialization of dideoxy DNA sequencing (Sanger, Nicklen, & Coulson, 1977) soon led to the identification of many putative P450-encoding genes. The first fully sequenced microbial genome, that of *Haemophilus influenza*, was solved by Fleischmann et al. (1995), with the number rising exponentially thereafter, reaching 2021 by 2006, 6285 by 2011, and over 32,500 at the time of writing this review. Today, coupled with ever more powerful bioinformatics tools for identification of novel genes of interest, there exists an exceptional resource for the discovery of P450s for both academic studies and commercial applications.

2. P450 CLASSES

Domain architecture and cellular location of the P450 and redox proteins categorize these systems into 1 of 10 defined classes (Hannemann et al., 2007). All except 2 of the 10 classes, Class IX and X, of P450 utilize a flavoprotein to facilitate electron transfer during catalysis. Such flavin-containing proteins are capable of being both oxidized and reduced; ie, they can accept and donate electrons and are thus employed in the transport of electrons from predominantly NAD(P)H ultimately to molecular oxygen.

Based on their ubiquity, Class I and II enzyme systems appear to offer great opportunity for research and industrial application, though other classes, such as the self-sufficient Class VII and VIII systems, offer particular advantages for commercial biocatalysis.

Class I systems (Fig. 4, Ia and Ib) employ a three-protein domain architecture, all of which are expressed from different genes but cofunction for enzyme activity. They have an FAD-dependent ferredoxin reductase, a ferredoxin containing an iron–sulfur cluster (1Fe–0S, 2Fe–2S, 3Fe–4S, 4Fe–3S, or 4Fe–4S), and a P450 constituent and require NAD(P)H as a source of electrons. This class contains most of the known bacterial P450s and also contains a number of eukaryotic mitochondrial systems (Fig. 4, Ia and Ib, respectively). Although members of the same class, these two subsystems are not closely phylogenetically related (Hannemann et al., 2007). In the bacterial type, all three components are soluble, whereas in the mitochondrial systems, only the ferredoxin is soluble and this shuttles electrons between membrane-associated ferredoxin reductase and P450 components. P450cam is the best-studied Class I bacterial system. It has a 2Fe–2S ferredoxin iron–sulfur cluster, referred to as putidaredoxin, and an FAD-dependent ferredoxin reductase, referred to as putidaredoxin reductase.

Class II systems (Fig. 5, II) employ a two-component system featuring a cytochrome P450 domain and a cytochrome P450 reductase (CPR) which binds FAD and FMN and requires NAD(P)H as a source of electrons. The CPR has evolved from a fusion of a 2Fe–2S ferredoxin with a ferredoxin reductase, where ferredoxin binding was lost in favor of FMN binding (Hannemann et al., 2007). This class of P450 system tends to exist most abundantly in the membranes of the endoplasmic reticulum of eukaryotes, with both domains apparently lying adjacent to one another. Some bacterial examples do exist, such as P450sca isolated from the soil bacterium *Streptomyces carbophilus* (Serizawa & Matsuoka, 1991), which was shown to catalyze the hydroxylation of mevastatin to pravastatin, a selective inhibitor involved

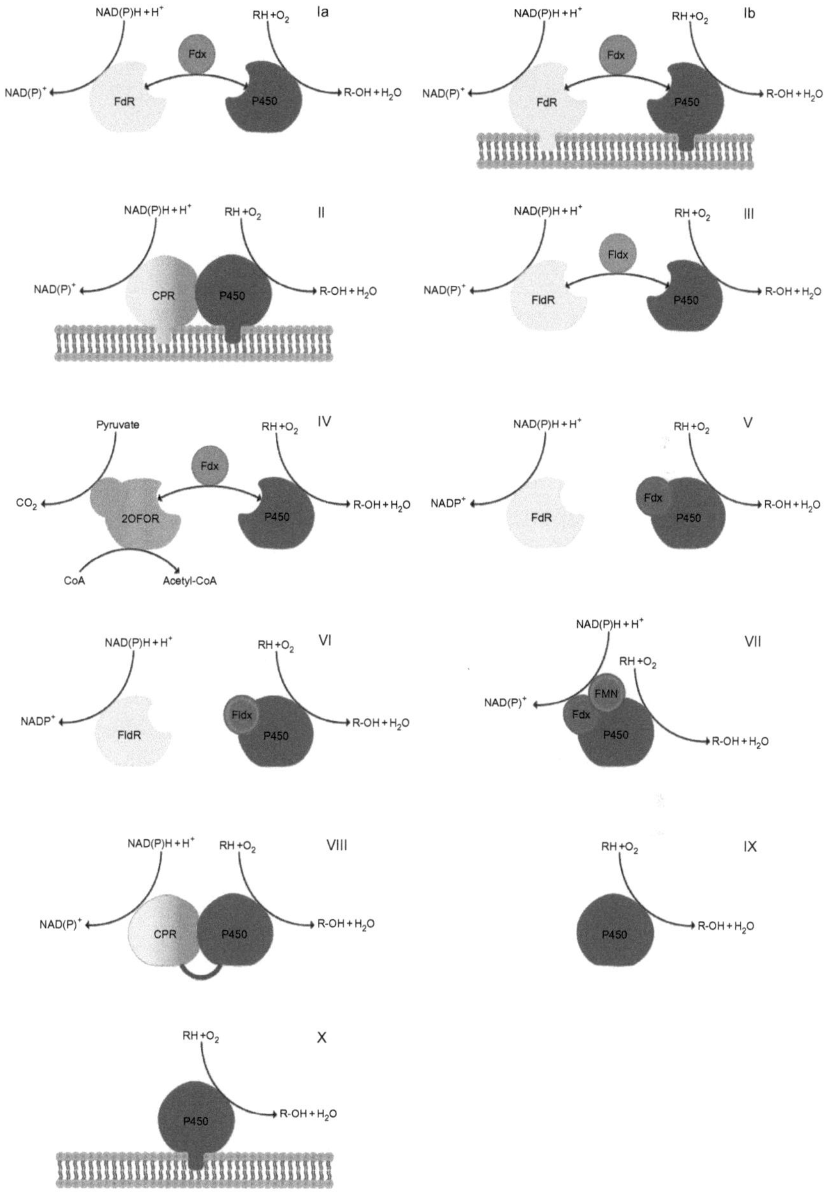

Fig. 4 Schematic outlining domain architecture of the different P450 systems.

in cholesterol biosynthesis. Due to the membrane bound nature of Class II enzyme systems, they are not generally regarded as possessing significant biocatalytic utility, due to likely difficulties in achieving sufficient recombinant expression, though they are extensively employed in drug metabolism

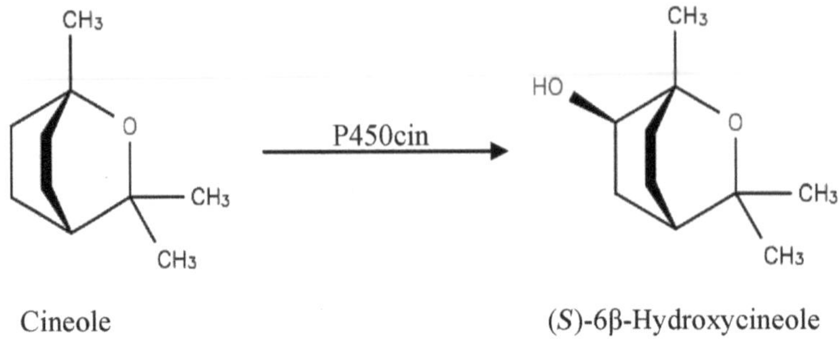

Fig. 5 Schematic outlining P450cin-mediated biotransformation of cineole to (*S*)-6β-hydroxycineole.

studies (Ingelman-Sundberg, 2004; Watanabe, Nara, & Serizawa, 1995; Zanger & Schwab, 2013).

Class III systems (Fig. 4, III) are much like bacterial Class I systems, in that they use three components comprising an FAD-dependent ferredoxin reductase, a small intermediate redox protein, and a P450. The difference, however, is that Class III systems utilize an FMN-dependent redox protein referred to as flavodoxin, rather than the iron–sulfur cluster containing Class I-type ferredoxin. This means that the Class III system is reminiscent of the Class II eukaryotic system which also utilizes FAD and FMN in its CPR electron transport protein, only that in the former these prosthetic groups reside within separate proteins. Very few enzymes have been attributed to this class to date, the most notable being P450cin, isolated from the bacterium *Citrobacter braakii* and named so for its activity toward the production of (*S*)-6β-hydroxycineole from cineole (Hawkes, Adams, Burlingame, Ortiz de Montellano, & De Voss, 2002; Fig. 5). The production of (*S*)-6β-hydroxycineole by P450cin allows *C. braakii* to utilize cineole as its sole carbon source (Meharenna, Slessor, Cavaignac, Poulos, & De Voss, 2008). Interestingly, *E. coli*, while not having an identified P450 catalyst, does have native flavodoxin and a flavodoxin reductase. Perhaps more significantly, these components have been shown in some instances to be capable of reducing heterologously expressed Class I P450s, such as P450biol from the bacterium *Bacillus subtilis*, which is involved in biotin synthesis (Lawson, von Wachenfeldt, Haq, Perkins, & Munro, 2004).

Class IV systems (Fig. 4, IV), of which few are currently known, reside exclusively in extremophiles. They are the only class of P450 which has been shown to utilize redox partners specific for an electron donor other than NAD(P)H. For example, CYP119A1, from *Sulfolobus solfataricus*, was observed to be reduced by a ferredoxin and a 2-oxo-acid:ferredoxin oxidoreductase originating in the same bacteria. This reductase system obtains electrons from pyruvic acid rather than NAD(P)H and the P450 has activity on lauric acid at an optimum temperature of 70°C (Puchkaev & Ortiz de Montellano, 2005).

Class V systems (Fig. 4, V) are represented by only a single characterized member to date, that of the sterol 14α-demethylase from *Methylococcus capsulatus*, which was shown to have a P450 domain bound to a 3Fe–4S-type ferredoxin at the C-terminus via an alanine-rich linker (Jackson et al., 2002).

Class VI systems (Fig. 4, VI) comprise a similar architecture to that of Class V, but the C-terminal-bound ferredoxin is replaced with an N-terminal-bound flavodoxin. The only example to date was identified in *Rhodococcus rhodochrous* by Seth-Smith et al. (2002). Both Class V and VI utilize an NAD(P)H-dependent ferredoxin reductase.

Class VII systems (Fig. 4, VII) present a completely novel domain architecture, where the P450 is C-terminally bound to a phthalate dioxygenase reductase domain, as suggested by sequence homology, ie, a "self-sufficient" P450. This type of reductase system comprises an iron–sulfur cluster and, unlike the FAD-dependent ferredoxin reductases, contains the prosthetic group FMN. The first enzyme of this kind was discovered by Roberts, Grogan, Greter, Flitsch, and Turner (2002) and named P450RhF. It was isolated from the bacterium *Rhodococcus* sp. strain NCIMB 9784 and shown to have hydroxylase activity against the organic fragrance compound coumarin in vivo, but intriguingly demonstrated only O-dealkylation activity in vitro.

Class VIII systems (Fig. 4, VIII), in common with Class VII, are also fully fused, self-sufficient P450s, where both systems possess an N-terminal P450 domain. However, whereas the P450 domain is followed by an FMN-containing reductase domain and then 2Fe–2S ferredoxin domain in Class VII enzymes, Class VIII enzymes instead employ an FMN-containing flavodoxin followed by an FAD-dependent C-terminal reductase. Self-sufficient P450s are significantly less ubiquitous than their Class I and II counterparts, but their single-gene nature makes them prime candidates for easier cloning and soluble expression. The archetype of Class VIII, P450BM3, from *Bacillus megaterium*, was discovered by Narhi and Fulco (1986) and has been extensively studied, especially from a mutagenesis point of view (Park et al., 2010; Reinen et al., 2011; Venkataraman et al., 2014). This enzyme is specific for medium- and long-chain fatty acids, both saturated and unsaturated. Its preferred substrate, arachidonic acid, a 20-carbon unsaturated fatty acid, is catalyzed at a rate (K_{cat}) of $\approx$17,000 min^{-1} (Munro,

Girvan, & McLean, 2007). It produces hydroxylated products at the ω-1, ω-2, and ω-3 positions, with no observable conversation at the terminal carbon.

Class IX systems (Fig. 4, IX) are unique in that they require no redox protein partners for transfer of electrons from NAD(P)H and consist solely of nitric oxide (NO) reductases. It is currently understood that Class IX catalysts do not possess monooxygenase activity. To date, they have only been identified in eukaryotes, typically fungi and yeast, with the first being discovered in the filamentous fungus *Fusarium oxysporum* (Kizawa et al., 1991). This archetypal enzyme was named P450nor. Since this discovery, a number of other P450nor-like enzymes have been investigated and genome-based analysis has shown the general presence of these catalysts in fungi (Daiber, Shoun, & Ullrich, 2005). It is now thought that these enzymes are involved in anaerobic respiration in fungi and yeast, catalyzing the transformation of NO to nitrous oxide (N_2O). An upregulation of the P450nor gene has even been observed in oxygen-limited conditions (Takaya et al., 1999).

Class X systems (Fig. 4, X) represent a very special case, as while being identified as "P450s," the members of this class are subdivided into a number of other types of enzyme class, namely hydroperoxide lyases (HPLs), allene oxide synthases (AOLs), divinyl ether synthases (DESs), prostacyclin synthases (PGISs), and thromboxane synthases (TBAXs). As with Class IX P450s, the members of Class X do not require the transport of electrons via redox proteins, but instead use differing intramolecular transfer systems. HPL, AOL, and DES enzymes have been shown to utilize the acyl hydroperoxide groups of their substrates as oxygen donors (Hannemann et al., 2007). To date, these P450s have only been found in plants and mammals.

3. P450 FAMILY NOMENCLATURE

P450s are also divided into families, each of which is based solely on amino acid sequence homology. Each family is given a number, each subfamily is given a letter, and then each gene within a subfamily is given a second number. This code is always prefixed with "CYP," denoting it as a cytochrome P450 (Nelson, 2006). P450cam for instance has the name CYP101A1, while P450BM3 has the name CYP102A1. Members of the same family are considered to have an amino acid sequence identity of greater than 40%, while members of the same subfamily have amino acid sequence homologies of greater than 55%.

It is important to note that an assignation of family identifies P450s with amino acid sequence identities within given percentage parameters, while that of class highlights the system by which a particular P450 gains the electrons required for catalysis. Such an observation meant it could be difficult to identify to which class a particular P450 being studied belonged. To address this, a study by Gricman, Vogel, and Pleiss (2014) involved the study of amino acid conservation across the different classes of P450. An alignment of 16,732 P450 genes and the study of a number of P450 crystal structures were performed. The group found a number of class-specific conserved residues, all of which lay on the proximal side of the P450, where interaction with the reductase proteins takes place. As an addition to their work, two "standard numbering" schemes were devised, to standardize residue positions across all P450s and thus aid in the consistency of work being published in the scientific literature. The group determined approximately 90% of all P450s belong to either Class I or Class II, and therefore each standard numbering scheme applied to one of those two classes. It was discovered that Class VII enzymes share the conservation of Class I P450s, while Class VIII enzymes share the conservation of Class II. As such, the Class I and II numbering schemes were applied to Class VII and VIII, respectively. Of the remaining six classes, too few sequences were known to deduce any conservation. A search tool, hosted on the Cytochrome P450 Engineering Database (CYPED; https://cyped.biocatnet.de/workbench/numbering) was also set up, to allow users to determine to which class any P450 belongs.

4. CATALYTIC CYCLE

The P450 mechanism employs a very complex, multistep catalytic cycle involving a range of transient intermediates. The enzymes mediate the catalytic oxidation of target compounds, where initially the substrate binds to the active site in the heme domain (Fig. 6, 2).

This prompts the release of a water molecule which covers the active site in the resting state of the enzyme. These reactions induce a conformational change in the shape of not only the active site, but with the self-sufficient catalysts (such as CYP102A1), also in the reductase domain, bringing the flavins closer together to facilitate the transport of electrons down the chain (Girvan et al., 2011). This results in transition of the iron from a low spin state to a high spin state (Mowat, Gazur, Campbell, & Chapman, 2010). The spin state simply refers to the energy associated with the iron and because of the binding substrate, the outer orbital has one or more unpaired

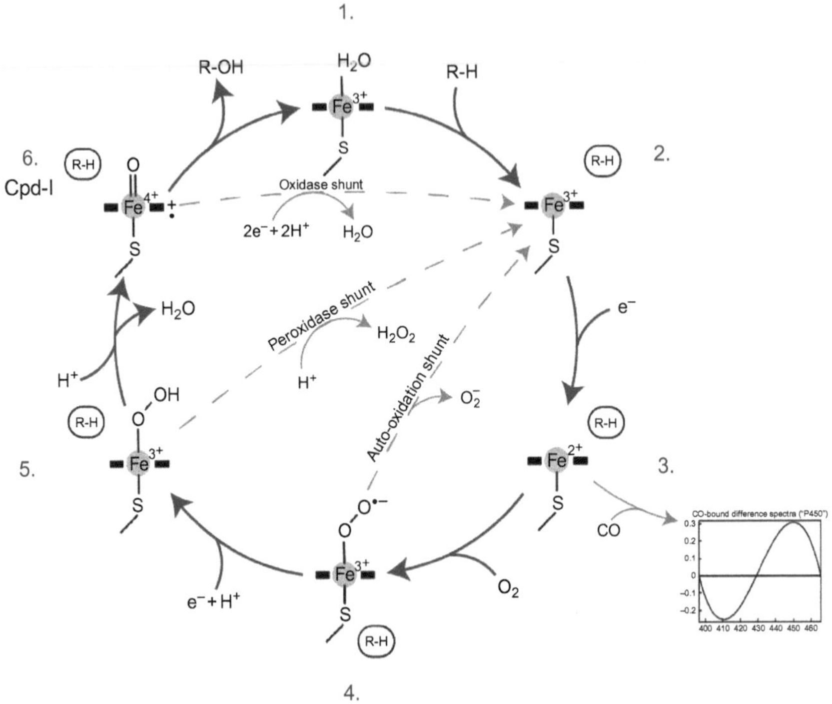

Fig. 6 Graphical representation of the catalytic cycle of the cytochromes P450.

electrons giving it a higher energy state, and thus it changes from a low to high spin. This change in energy is also the reason for the commonly noted "difference spectra" observed when the enzyme substrate complex is measured spectrophotometrically. The difference spectrum is essentially two shifts in the complex's absorbance; an increase in absorbance at 390 nm is observed, coupled with a decrease in absorbance at 420 nm. Alternatively, if carbon monoxide adducts to a reduced P450, an absorbance shift to the well noted 450 nm occurs. The conformational change during substrate binding instigates the primary electron donation in the presence of NAD(P)H. An electron passes from NAD(P)H to the first flavin molecule, FAD in the case of P450BM3, then onto FMN, and finally to the heme group where it is accepted by the iron at the porphyrin ring. This changes iron from its ferric state (Fe^{3+}) to its ferrous state (Fe^{2+}) (Fig. 6, 3). Once the initial reduction has occurred, molecular oxygen will then bind covalently at the distal axial position of the heme iron and activate the oxygen (Munro et al., 2007) (Fig. 6, 4). Dissociation of the iron–oxygen bond can occur at this stage creating a superoxide radical, which is very reactive and can disrupt the catalytic cycle, described in greater detail later. The second reduction step then occurs, which results in the formation of a negatively charged peroxo group, where the iron-bound oxygen becomes more negatively charged than in its previous state prior to electron donation (Fig. 6, 5). This peroxo group is very nucleophilic, making this intermediate stage very short lived. The complex is then rapidly protonated twice due to hydrogen binding from surrounding water and amino acid side chains. This results in the scission of the dioxygen which binds to the donated hydrogen giving rise to the release of a water molecule and the formation of the cytochrome P450 compound I intermediate (Fig. 6, 6). This intermediate has an Fe^{4+} group due to the donated electron pair from the nucleophilic peroxo group and has a double-bonded oxygen atom. Cytochrome P450 compound I (Cpd-I) is widely accepted as being the final intermediate stage in the P450 catalytic cycle and is thought to be the main oxidizing agent which creates the hydroxylated product (Rittle, Younker, & Green, 2010). Cpd-I has a delocalized oxidizing equivalent over the porphyrin ring and C–S bond within the heme domain making it highly reactive. Cpd-I oxidizes the bound substrate which is subsequently released as hydroxylated product. Water then binds within the active site, returning the enzyme to its original resting state (Fig. 6, 1).

5. UNCOUPLING

As mentioned previously, once molecular oxygen has bound to the heme iron, dissociation can occur, resulting in the cessation of catalysis. This can occur in one of the three ways: (i) via the auto-oxidation shunt pathway, where the anion superoxide radical is formed after degradation of the peroxo-ferrous intermediate, (ii) via the peroxidase shunt pathway, where protonation of the hydroxyferric intermediate leads to dissociation and the release of hydrogen peroxide, or (iii) via the oxidase shunt pathway, where a water molecule is released after the double reduction and diprotonation of Cpd-I (Fig. 6).

An instance of any of these shunt pathways occurring is known as "uncoupling," with the coupling efficiency of a P450 being an important determinant of the enzymes rate of catalysis. The coupling efficiency is a measure of the number of electrons donated from NAD(P)H which are successfully used for the catalysis of a bound organic substrate. P450cam for instance has a coupling efficiency against camphor of 100%, meaning that all electrons donated from the nicotinamide cofactor are used for catalysis to 5-hydroxycamphor. The same enzyme however has a coupling efficiency against styrene of only 2–7% (Nickerson, Harford-Cross, Fulcher, & Wong, 1997). This observation suggests that the mechanics of substrate binding has a profound effect on the flow of electrons to the heme center, greatly influencing the rate of catalytic turnover. As may be expected, great attention has been focused on improving the coupling efficiency of P450 enzymes for nonnative substrates. A number of important residues have been identified which surround the heme domain, which appear to play a crucial role in the efficient delivery of electrons for the scission of dioxygen. One such study by Yoshioka, Takahashi, Ishimori, and Morishima (2000) demonstrated the importance of a leucine residue in the "cysteine pocket" of P450cam. When mutated to a proline, the uncoupling rates against a number of nonnative substrates were improved and resulted in significantly increased rates of activity. In addition to this, activity on the native substrate, camphor, was retained, albeit with a decrease in the activity of approximately 12%, but with the coupling efficiency remaining at 100%. It was concluded that the mutation had removed a conserved amide proton, which had enhanced what they termed the "push effect," an effect where upon force can be synergistically applied from surrounding hydrogen bonds to aid in cleaving dioxygen (Ogliaro, de Visser, & Shaik, 2002). This provided more favorable production of Cpd-I, the final intermediate in the production of oxidized product (Fig. 6).

Ost et al. (2003) demonstrated the importance of a highly conserved phenylalanine residue which lies in close proximity to the cysteine of the thiolate ligand. In P450BM3, this residue lies at position 393 and was the focus of their study. The phenylalanine was mutated to a number of different amino acids, namely Ala, His, Tyr, and Trp. All four mutants expressed in soluble form and were shown to fold correctly, displaying the typical reduced, carbon monoxide-bound absorption spectra. Interestingly, however, each mutant had an altered reduction potential, with ranges above and below that of the wild type. The effect this had on catalytic turnover only ever showed a decrease in rate, suggesting that the reduction potential of P450BM3 has been naturally optimized for the most efficient rate of product formation. Despite a decrease in turnover number, three of the four mutants showed a decreased rate of uncoupling, with the F393H mutant showing an excess of 40-fold lower rate of uncoupling by comparison to the WT enzyme. This would suggest that the flow of electrons had become rate limiting, as both creation of product and superoxide radicals were limited.

Another study, by Matsumura et al. (2008), identified two key residues in P450st, a thermophilic catalyst from the organism *Sulfolobus tokodaii*. The mutations selected were on the basis of conservation within the binding motif of P450s and thus were comparable with mutants produced in previous studies with both P450cam and P450BM3. The mutants F310A and A320Q were produced along with the double mutant F310A/A320Q, with both residues being in close proximity to the thiolate ligand of the heme. All three showed an increased reduction potential, allowing for a better flow of electrons to the heme center and as such each mutant demonstrated a greater turnover number for activity against styrene.

6. P450s IN INDUSTRY

As can be seen from the complexity of the P450 pathway and the requirement for a number of partner proteins and cofactors, it is difficult to isolate and produce P450s which can be utilized in a commercial setting. Other issues such as low stability and poor turnover numbers add to this. Despite these issues, the extraordinary capability of this family of enzymes makes them very exciting targets for things such as drug production, flavor compound production, and pesticide production. As such, a wide reaching effort has been made to both isolate and manipulate P450s into effective catalysts for a number of commercial applications.

One of the most notable examples of the few industrial applications of a P450 is the 6β-hydroxylation of compactin, a statin isolated from the mesophilic fungus *Penicillium citrinum*. Otherwise known as mevastatin, this compound went through clinical trials but was never marketed. It was instead used as a precursor in the production of pravastatin (Fig. 7; Watanabe et al., 1995).

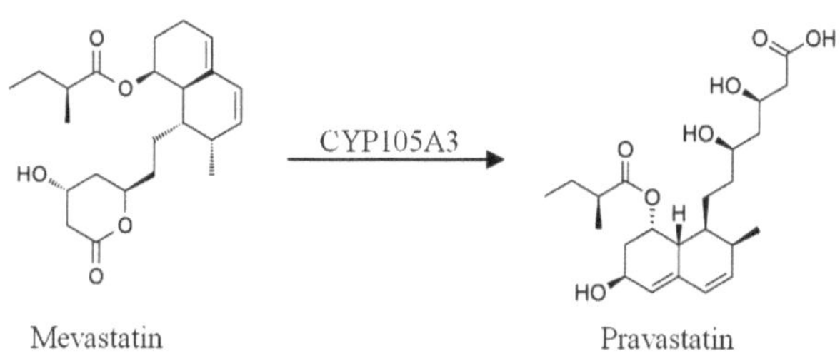

Fig. 7 Schematic outlining CYP105A3-mediated biotransformation of mevastatin to pravastatin.

Pravastatin is a low-density lipoprotein, molecules which aid in the transport of fats around the body. It is very potent in treating patients who are hyperlipidemic or, more simply, have high levels of lipids in the blood. It is a drug which is used in over 100 countries, with the industrial scale production being developed by Sankyo Co., Ltd. in Japan. It is produced via the whole-cell biotransformation of compactin in the organism *S. carbophilus*, hyper-expressing the P450sca-2 (CYP105A3) gene.

Another noteworthy example is the production of 1α,25-dihydroxyvitamin D_3 from vitamin D_3, performed by a P450 from the CYP105 family. This enzyme catalyzes a two-stage reaction, the biotransformation of vitamin D_3 at the C-25 position to produce 25-hydroxyvitamin D_3 and then the hydroxylation of this intermediate at the C-1-alpha position to produce 1α,25-dihydroxyvitamin D_3 (Fig. 8; Sugimoto et al., 2008).

It is the twice hydroxylated form of vitamin D_3 which is physiologically active in humans and is used to treat a number of conditions, namely: hyperthyroidism, osteoporosis, and renal failure. Other P450s however have shown similar activity, like a P450 from the bacterium *Pseudonocadria autotrophica*. This enzyme, often known as Vdh, is from the CYP107 family of P450s and has been manipulated extensively in an effort to improve its activity, with some success. One particular effort, by Hayashi et al. (2010), resulted in the production of a double mutant with a 400-fold increase in K_{cat}. The bioconversion of vitamin D_3 is a prime example of the use of P450s in green chemistry. The previous method of chemical conversion was a 20-stage process and could often result in product yields of as little as 1% (Zhu & Okamure, 1995).

One further example of a commercially applied P450, CYP108B1, showed activity against terpenes, chemicals often used as flavor compounds and fragrances. Usually found in grapefruit, (+)-nootkatone is a high value flavor compound which is difficult to make chemically. Girhard et al. (2009)

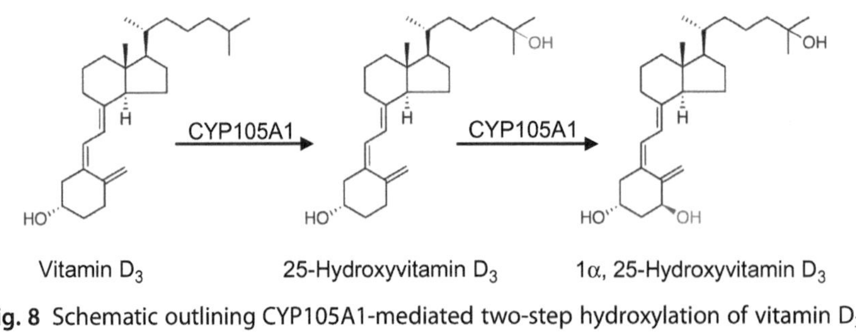

Fig. 8 Schematic outlining CYP105A1-mediated two-step hydroxylation of vitamin D_3.

utilized a high-throughput screening approach of over 100 P450s in an effort to produce (+)-nootkatone from its precursor molecule (+)-valencene. The enzymes were screened in both whole-cell and cell-free formats and were supplemented with the P450cam reductase proteins putidaredoxin and putidaredoxin reductase as electron transfer partners. Using CYP108B1 in cell-free format, a number of by-products were observed, accounting for over 30% of the product formed. Eventually utilizing a two-phase system with water and 10% dodecane, the group were able to achieve yields of up to 120 mg/L within 8 h, with (+)-nootkatone accounting for over 97% of the product produced.

Girhard et al. (2009) proposed the question "what is the best route to discovery of putative P450 enzymes?" Mutation or high-throughput screening of enzymes from nature. Both approaches have complementary and varying difficulties. For both, perhaps the main barrier to discovery is finding redox partner proteins which can be used to transfer electrons from the nicotinamide cofactor to the heme center. While the number of P450s within the superfamily is growing exponentially with the advancement of genomic sequencing, the discovery of redox partner proteins of the nonself-sufficient classes has not progressed as promisingly. This is largely due to a lack of colocation of the redox partner and P450 genes within the genome. With that said, a number of works have suggested the ability of cross-reactivity between P450s and nonnative reductase systems, such as the conversion of (+)-valencene (Girhard et al., 2009). Whether the activity achieved is representative of the P450s true potential with its native system is yet to be seen and is highly unlikely. Another example would be a study by Scheps, Malca, Hoffman, Nestl, and Hauer (2011) where the group demonstrated cross-reactivity of the P450cam reductase system proteins, putidaredoxin and putidaredoxin reductase with two Class I P450 enzymes from the CYP153 superfamily. They were able to hydroxylate a range of short- to medium-chain alkanes to produce the corresponding alcohols and diols. By their own admission the relative product yields were quite low, with the production of 1-hydroxyoctane being the greatest at 165 μM final concentration after 4 h incubation in vitro. Despite this, it did further demonstrate the ability of P450s to overcome the barrier of electron transport and highlighted the real potential these enzymes possess.

Another widely explored method aimed at modifying P450s is that of tethering and domain switching. This method attempts to incorporate short linker regions between the domains of mostly Class I P450s and nonnative reductase systems, in an attempt to enhance electron transport. It is an approach mostly targeted at, but not limited to, the Class VIII enzyme systems, such as P450BM3. The unique single protein architecture of these enzyme systems makes them ideal candidates for manipulation studies, mainly with respect to switching the P450 domain for a nonnative target with the aim of creating a new single gene, chimeric catalyst.

One domain switching study, carried out by Dodhia, Fantuzzi, and Gilardi (2006), resulted in the production and soluble expression of three active chimeric catalysts, utilizing the P450BM3 reductase system fused to three different human P450s. The group were able to demonstrate the ability of these chimeras to catalyze their already established reactions: diclofenac hydroxylation (CYP2C9), omeprazole hydroxylation (CYP2C19), and erythromycin N-demethylation (CYP3A4). In addition, the chimeras could achieve similar rates of activity and coupling efficiencies to those which were observed with their native reductase systems. This is significant, as not only have the two fused components come from different classes of P450 but also from different organisms, hinting at the possible scope of this approach to create self-sufficient catalysts which could ultimately be used in industry.

A further study utilized the reductase system from the Class VII P450RhF, which was fused, using a short linker, to the P450 domain of P450cam (CYP101A1; Robin et al., 2011). The group demonstrated the chimeras ability to hydroxylate 30 mM of (+)-camphor, the preferred substrate of CYP101A1, in whole-cell format in 48 h, with a final yield of 80%. In addition to this, the group demonstrated the importance of the length of the linker region in catalysis, along with the pH of the buffer and reaction temperature. The final yield was achieved with a linker length of nine amino acids, at pH 7.0 with a reaction temperature of 4°C.

Genomic data suggest that many organisms have and unequal number of ferredoxin- and reductase-encoding genes compared to that of P450s. A good example of this is *E. coli*, which as mentioned previously, has no identifiable P450 gene, yet it has as many as seven putidaredoxin like ferredoxin genes and at least one NAD(P)H reductase gene (Pfam protein family database entries, PF14759 and PF00111). This would suggest that the reductase system proteins service more than the P450 family. Therefore, the elucidation of native reductase-P450 pairs can be difficult, as potentially huge numbers of redox proteins would need to be cloned to match up to a large library of P450s. It would also suggest that a single NAD(P)H reductase enzyme is capable of reducing more than one ferredoxin. With this in mind, the ability of P450s to cooperate with nonnative redox systems becomes very important, as if one reductase could reduce a library of ferredoxins, perhaps a small collection of both proteins would be enough to find electron-donating proteins capable of activating a range of P450s.

A problem unique to high-throughput screening is the approach taken for target selection. There are many classes and over 114,000 P450 sequences known (Pfam PF00067), making rational and diverse selection of targets from this vast resource overwhelming.

While mutation offers a valuable tool toward the creation of novel P450 enzymes, it is often difficult to produce mutants rationally, due to the limited knowledge we have of these catalysts. As previously outlined, many studies have used this approach, and some with success, but a less rational approach can be time consuming and expensive with little chance of discovery.

REFERENCES

Bernhardt, R. (2006). Cytochromes P450 as versatile biocatalysts. *Journal of Biotechnology, 124*(1), 128–145.

Brodie, B. B., Axelrod, J., Cooper, J. R., Gaudette, L., La Du, B. N., Mitoma, C., et al. (1955). Detoxication of drugs and other foreign compounds by liver microsomes. *Science, 121*, 603–604.

Daiber, A., Shoun, H., & Ullrich, V. (2005). Nitric oxide reductase (P450nor) from *Fusarium oxysporum*. *Journal of Inorganic Biochemistry, 99*(1), 185–193.

Dodhia, V. R., Fantuzzi, A., & Gilardi, G. (2006). Engineering human cytochrome P450 enzymes into catalytically self-sufficient chimeras using molecular Lego. *Journal of Biological Inorganic Chemistry, 11*(7), 903–916.

Fleischmann, R. D., Adams, M. D., White, O., Clayton, R. A., Kirkness, E. F., Kerlavage, A. R., et al. (1995). Whole-genome random sequencing and assembly of *Haemophilus influenzae* Rd. *Science, 269*, 496–512.

Fujii, T., Narikawa, T., Sumisa, F., Arisawa, A., Takeda, K., & Kato, J. (2006). Production of alpha, omega-alkanediols using *Escherichia coli* expressing cytochrome P450 from *Acinetobacter* sp. OC4. *Bioscience, Biotechnology, and Biochemistry, 70*(6), 1379–1385.

Girhard, M., Machida, K., Itoh, M., Schmid, R. D., Arisawa, A., & Urlacher, V. B. (2009). Regioselective biooxidation of (+)-valencene by recombinant *E. coli* expressing CYP109B1 from *Bacillus subtilis* in a two-liquid-phase system. *Microbial Cell Factories, 8*, 36.

Girvan, H. M., Dunford, A. J., Neeli, R., Ekanem, I. S., Waltham, T. N., Joyce, M. G., et al. (2011). Flavocytochrome P450 BM3 mutant W1046A is a NADH-dependent fatty acid hydroxylase: Implications for the mechanism of electron transfer in the P450 BM3 dimer. *Archives of Biochemistry and Biophysics, 507*(1), 75–85.

Gricman, L., Vogel, C., & Pleiss, J. (2014). Conservation analysis of class-specific positions in cytochrome P450 monooxygenases: Functional and structural relevance. *Proteins, 82*(3), 491–504.

Grogan, G., Roberts, S., Wan, P., & Willetts, A. (1993). Camphor-grown *Pseudomonas putida*, a multifunctional biocatalyst for undertaking Baeyer-Villiger monooxygenase dependent biotransformations. *Biotechnology Letters, 15*(9), 913–918.

Guengerich, P. F., & Cheng, Q. (2011). Orphans in the human cytochrome P450 superfamily: Approaches to discovering functions and relevance in pharmacology. *Pharmacological Reviews, 63*(3), 684–699.

Hannemann, F., Bichet, A., Ewen, K. M., & Bernhardt, R. (2007). Cytochrome P450 systems—Biological variations of electron transport chains. *Biochimica et Biophysica Acta, 1770*, 330–344.

Hawkes, D. B., Adams, G. W., Burlingame, A. L., Ortiz de Montellano, P. R., & De Voss, J. J. (2002). Cytochrome P450(cin) (CYP176A), isolation, expression, and characterization. *The Journal of Biological Chemistry, 277*(31), 27725–27732.

Hayashi, K., Yasuda, K., Sugimoto, H., Ikushiro, S., Kamakura, M., Kittaka, A., et al. (2010). Three-step hydroxylation of vitamin D3 by a genetically engineered CYP105A1: Enzymes and catalysis. *The FEBS Journal, 277*(19), 3999–4009.

Ingelman-Sundberg, M. (2004). Human drug metabolising cytochrome P450 enzymes: Properties and polymorphisms. *Naunyn-Schmiedeberg's Archives of Pharmacology, 369*(1), 89–104.

Jackson, C. J., Lamb, D. C., Marczylo, T. H., Warrilow, A. G., Manning, N. J., Lowe, D. J., et al. (2002). A novel sterol 14alpha-demethylase/ferredoxin fusion protein (MCCYP51FX) from *Methylococcus capsulatus* represents a new class of the cytochrome P450 superfamily. *The Journal of Biological Chemistry, 277*(49), 46959–46965.

Katagiri, M., Ganguli, B. N., & Gunsalus, I. (1968). A soluble cytochrome P-450 functional in methylene hydroxylation. *The Journal of Biological Chemistry, 243*(12), 3543–3546.

Kizawa, H., Tomure, D., Oda, M., Fukamizu, A., Hoshimo, T., Gotoh, O., et al. (1991). Nucleotide sequence of the unique nitrate/nitrite-inducible cytochrome P-450 cDNA from *Fusarium oxysporum*. *The Journal of Biological Chemistry, 266*(16), 10632–10637.

Klingenberg, M. (1958). Pigments of rat liver microsomes. *Archives of Biochemistry and Biophysics, 75*(2), 376–386.

Lawson, R. J., von Wachenfeldt, C., Haq, I., Perkins, J., & Munro, A. (2004). Expression and characterization of the two flavodoxin proteins of *Bacillus subtilis*, YkuN and YkuP: Biophysical properties and interactions with cytochrome P450 bioI. *Biochemistry, 43*, 12390–12409.

Matsumura, H., Wakatabi, M., Omi, S., Ohtaki, A., Makamura, N., Yohda, M., et al. (2008). Modulation of redox potential and alteration in reactivity via the peroxide shunt pathway by mutation of cytochrome P450 around the proximal heme ligand. *Biochemistry, 47*, 4834–4841.

Meharenna, Y. T., Slessor, K. E., Cavaignac, S. M., Poulos, T. L., & De Voss, J. J. (2008). The critical role of substrate-protein hydrogen bonding in the control of regioselective hydroxylation in p450cin. *The Journal of Biological Chemistry, 283*(16), 10804–10812.

Mowat, C. G., Gazur, B., Campbell, L. P., & Chapman, S. K. (2010). Flavin-containing heme enzymes. *Archives of Biochemistry and Biophysics, 493*(1), 37–52.

Munro, A. W., Girvan, H. M., & McLean, K. J. (2007). Cytochrome P450—Redox partner fusion enzymes. *Biochimica et Biophysica Acta, 1770*(3), 345–359.

Narhi, L. O., & Fulco, A. J. (1986). Characterization of a catalytically self-sufficient 119,000-Dalton cytochrome P-450 monooxygenase induced by barbiturates in *Bacillus megaterium*. *The Journal of Biological Chemistry, 261*(16), 7160–7169.

Nelson, D. R. (2006). Cytochrome P450 nomenclature, 2004. *Methods in Molecular Biology, 320*, 1–10.

Nickerson, D. P., Harford-Cross, C. F., Fulcher, S. R., & Wong, L.-L. (1997). The catalytic activity of cytochrome P450cam towards styrene oxidation is increased by site-specific mutagenesis. *FEBS Letters, 405*(2), 153–156.

Niwa, T., Murayama, N., & Yamazaki, H. (2011). Stereoselectivity of human cytochrome p450 in metabolic and inhibitory activities. *Current Drug Metabolism, 12*(6), 549–569.

Oezguen, N., Kumar, S., Hindupur, A., Braun, W., Muralidhara, B. K., & Halpert, J. R. (2008). Identification and analysis of conserved sequence motifs in cytochrome P450 family 2. Functional and structural role of a motif 187RFDYKD192 in CYP2B enzymes. *The Journal of Biological Chemistry, 283*(31), 21808–21816.

Ogliaro, F., de Visser, S. P., & Shaik, S. (2002). The 'push' effect of the thiolate ligand in cytochrome P450: A theoretical gauging. *Journal of Inorganic Chemistry, 91*, 554–567.

Omura, T., & Sato, R. (1964). The carbon monoxide-binding pigment of liver microsomes: I. Evidence for its hemoprotein nature. *Journal of Biological Chemistry, 239*(7), 2370–2378.

Ost, T. W. B., Clark, J., Mowat, C., Miles, C. S., Walkinshaw, M. D., Reid, G. A., et al. (2003). Oxygen activation and electron transfer in flavocytochrome P450 BM3. *Journal of the American Chemical Society, 125*, 15010–15020.

Park, S. H., Kim, D. H., Kim, D., Kim, D. H., Jung, H. C., Pan, J. G., et al. (2010). Engineering bacterial cytochrome P450 (P450) BM3 into a prototype with human P450 enzyme activity using indigo formation. *Drug Metabolism and Disposition, 38*(5), 732–739.

Podust, L. M., & Sherman, D. H. (2012). Diversity of P450 enzymes in the biosynthesis of natural products. *Natural Product Reports, 29*(10), 1251–1266.

Poulos, T. L., Finzel, B. C., Gunsalusll, I. C., Wagner, G., & Kraut, J. (1985). The 2.6-Å crystal structure of *Pseudomonas putida* cytochrome P-450. *The Journal of Biological Chemistry, 260*(30), 16122–16130.

Puchkaev, A. V., & Ortiz de Montellano, P. R. (2005). The *Sulfolobus solfataricus* electron donor partners of thermophilic CYP119: An unusual non-NAD(P)H-dependent cytochrome P450 system. *Archives of Biochemistry and Biophysics, 434*(1), 169–177.

Raucy, J. L., & Allen, S. W. (2001). Recent advances in P450 research. *Pharmacogenetics, 1*, 178–186.

Reinen, J., van Leeuwen, J. S., Li, Y., Sun, L., Grootenhuis, P. D., Decker, C. J., et al. (2011). Efficient screening of cytochrome P450 BM3 mutants for their metabolic activity and diversity toward a wide set of drug-like molecules in chemical space. *Drug Metabolism and Disposition, 39*(9), 1568–1576.

Rittle, J., Younker, J. M., & Green, M. T. (2010). Cytochrome P450: The active oxidant and its spectrum. *Inorganic Chemistry, 49*(8), 3610–3617.

Roberts, G. A., Grogan, G., Greter, A., Flitsch, S. L., & Turner, N. J. (2002). Identification of a new class of cytochrome P450 from a Rhodococcus sp. *Journal of Bacteriology, 184*(14), 3898–3908.

Robin, A., Kohler, V., Jones, A., Ali, A., Kelly, P. P., O'Reilly, E., et al. (2011). Chimeric self-sufficient P450cam-RhFRed biocatalysts with broad substrate scope. *Beilstein Journal of Organic Chemistry*, 7, 1494–1498.

Rupasinghe, S., Schuler, M. A., Kagawa, N., Yuan, H., Lei, L., Zhao, B., et al. (2006). The cytochrome P450 gene family CYP157 does not contain EXXR in the K-helix reducing the absolute conserved P450 residues to a single cysteine. *FEBS Letters, 580*(27), 6338–6342.

Sanger, F., Nicklen, S., & Coulson, R. (1977). DNA sequencing with chain-terminating inhibitors. *Proceedings of the National Academy of Sciences of the United States of America, 17*(12), 5463–5467.

Scheps, A., Malca, S. H., Hoffman, H., Nestl, B. M., & Hauer, B. (2011). Regioselective ω-hydroxylation of medium-chain *n*-alkanes and primary alcohols by CYP153 enzymes from *Mycobacterium marinum* and *Polaromonas* sp. strain JS666. *Organic and Biomolecular Chemistry, 9*, 6727–6733.

Serizawa, N., & Matsuoka, T. (1991). A two component-type cytochrome P-450 monooxygenase system in a prokaryote that catalyzes hydroxylation of ML-236B to pravastatin, a tissue-selective inhibitor of 3-hydroxy-3-methylglutaryl coenzyme A reductase. *Biochimica et Biophysica Acta, 1084*, 35–40.

Seth-Smith, H. M. B., Rosser, S. J., Basran, A., Travis, E. R., Dabbs, E. R., Nicklin, S., et al. (2002). Cloning, sequencing, and characterization of the hexahydro-1,3,5-trinitro-1,3,5-triazine degradation gene cluster from *Rhodococcus rhodochrous*. *Applied and Environmental Microbiology, 68*(10), 4764–4771.

Sugimoto, H., Shinkyo, R., Hayashi, K., Yoneda, S., Yamada, M., Kamakura, M., et al. (2008). Crystal structure of CYP105A1 (P450SU-1) in complex with 1α,25-dihydroxyvitamin D3. *Biochemistry*, *47*, 4017–4027.

Takaya, N., Suzuki, S., Kuwazaki, S., Shoun, H., Maruo, F., Yamaguchi, M., et al. (1999). Cytochrome P450nor, a novel class of mitochondrial cytochrome P450 involved in nitrate respiration in the fungus *Fusarium oxysporum*. *Archives of Biochemistry and Biophysics*, *372*(2), 340–346.

Urlacher, V. B. (2010). Catalysis with cytochrome P450 monooxygenases. *Handbook of Green Chemistry*, *3*(1), 1–25.

Venkataraman, H., Verkade-Vreeker, M. C., Capoferri, L., Geerke, D. P., Vermeulen, N. P., & Commandeur, J. N. (2014). Application of engineered cytochrome P450 mutants as biocatalysts for the synthesis of benzylic and aromatic metabolites of fenamic acid NSAIDs. *Bioorganic and Medicinal Chemistry*, *22*(20), 5613–5620.

Watanabe, I., Nara, F., & Serizawa, N. (1995). Cloning, characterization and expression of the gene encoding cytochrome P-450sca_2 from Streptomyces carbophilus involved in production of pravastatin, a specific HMG-CoA reductase inhibitor. *Gene*, *163*, 81–85.

Yoshioka, S., Takahashi, S., Ishimori, K., & Morishima, I. J. (2000). Roles of the axial push effect in cytochrome P450cam studied with the site-directed mutagenesis at the heme proximal site. *Journal of Inorganic Biochemistry*, *81*(3), 141–151.

Zanger, U. M., & Schwab, M. (2013). Cytochrome P450 enzymes in drug metabolism: Regulation of gene expression, enzyme activities, and impact of genetic variation. *Pharmacology and Therapeutics*, *138*(1), 103–141.

Zhu, G. D., & Okamure, W. H. (1995). Synthesis of vitamin D (Calciferol). *Chemical Reviews*, *95*, 1877–1952.

CHAPTER SIX

Understanding the Multifaceted Role of Human Down Syndrome Kinase DYRK1A

L.J. Kay, T.K. Smulders-Srinivasan, M. Soundararajan[1]
Faculty of Health and Life Sciences, Northumbria University, Newcastle upon Tyne, United Kingdom
[1]Corresponding author: e-mail address: meera.soundararajan@northumbria.ac.uk

Contents

1. Introduction 128
2. Minibrain in *Drosophila* 129
3. Role of DYRK1A in Infection and Immunity 132
4. Cardiovascular Function of DYRK1A 136
5. Role of DYRK1A in Cancer 139
   5.1 DYRK1A and the p53 Family 139
   5.2 DYRK1A in Apoptosis 141
   5.3 DYRK1A as a Promoter of Cell Proliferation 145
   5.4 DYRK1A in Myeloproliferative Disorders 147
6. Understanding the Neuronal Function of DYRK1A 148
   6.1 DYRK1A in Neuronal Function 149
   6.2 DYRK1A in Tau-Related Neurodegenerative Diseases 149
   6.3 Tau Splicing Regulation by DYRK1A 151
   6.4 DYRK1A Is a Synuclein Regulator 153
   6.5 Role of DYRK1A in Parkinson's Disease (PD) 155
7. Concluding Remarks 157
References 158

Abstract

The dual-specificity tyrosine (Y) phosphorylation-regulated kinase DYRK1A, also known as Down syndrome (DS) kinase, is a dosage-dependent signaling kinase that was originally shown to be highly expressed in DS patients as a consequence of trisomy 21.

Although this was evident some time ago, it is only in recent investigations that the molecular roles of DYRK1A in a wide range of cellular processes are becoming increasingly apparent. Since initial knowledge on DYRK1A became evident through minibrain *mnb*, the *Drosophila* homolog of DYRK1A, this review will first summarize the scientific reports on minibrain and further expand on the well-established neuronal functions of mammalian and human DYRK1A. Recent investigations across the current decade have provided rather interesting and compelling evidence in establishing nonneuronal

Advances in Protein Chemistry and Structural Biology, Volume 105
ISSN 1876-1623
http://dx.doi.org/10.1016/bs.apcsb.2016.07.001

functions for DYRK1A, including its role in infection, immunity, cardiomyocyte biology, cancer, and cell cycle control. The latter part of this review will therefore focus in detail on the emerging nonneuronal functions of DYRK1A and summarize the regulatory role of DYRK1A in controlling Tau and α-synuclein. Finally, the emerging role of DYRK1A in Parkinson's disease will be outlined.

1. INTRODUCTION

The dual-specificity tyrosine phosphorylation-regulated kinase DYRK1A belongs to the DYRK subfamily, a subgroup of the larger CMGC family of protein kinases (Becker & Joost, 1999; Becker et al., 1998). The biological role of DYRK1A was first brought to light from mapping of the gene to a critical region of chromosome 21; human trisomy of this chromosome results in Down syndrome (DS), one of the most common congenital abnormalities found in humans. DYRK1A is located in a region termed the "Down syndrome critical region" (DSCR), which is sufficient to produce neuronal abnormalities if present in three copies. It is important to note that DS is a multifactorial disease and that individuals harboring this syndrome suffer from a range of abnormalities including cardiac failure, musculoskeletal abnormalities, and inadequate immune responses in addition to the neuronal features initially reported. Interestingly, the symptoms associated with DS vary widely from mild to very severe and so accordingly do the associated clinical features. It has now been well established that children with DS have a higher risk of leukemia and remarkably increased occurrence of acute lymphoblastic leukemia (ALL) and acute myeloid leukemia compared to normal children. Interestingly, this situation reverses among DS adults, where a reduced occurrence of any epithelial cancer is seen relative to control cohorts.

Human DYRK1A has been previously demonstrated as a unique protein kinase capable of autophosphorylation on tyrosine residues and phosphorylating the substrates on serine/threonine (Alvarez, Altafaj, Aranda, & de la Luna, 2007; Himpel et al., 2001). Initial studies showed the transitional autophosphorylation of the critical activation-loop tyrosine in DYRK1A is intramolecular and mediated by the nascent kinase passing through a transitory intermediate form, suggesting a unique mode of regulation and constitutively active DYRK1A (Lochhead, Sibbet, Morrice, & Cleghon, 2005). However, later studies clearly established that the activity of DYRK1A is regulated intramolecularly, through Dyrk homology (DH) and NAPA

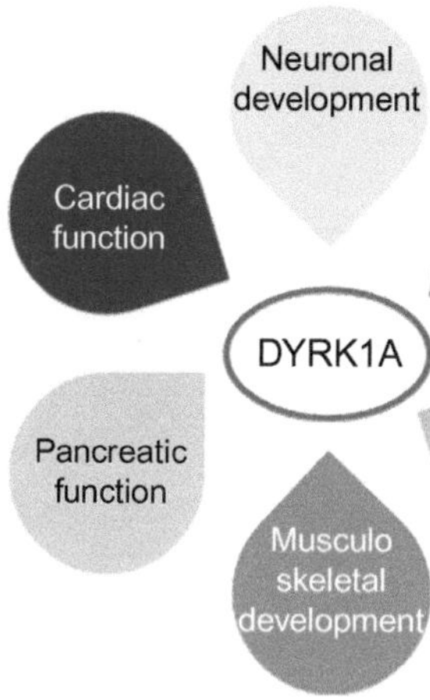

Fig. 1 Multifunctional role of DYRK1A. The central role of DYRK1A in controlling various physiological processes is outlined using the cartoon above.

regions, and possibly also through the 14-3-3-binding site in the C-terminus of the molecule (Alvarez et al., 2007; Soundararajan et al., 2013) (see Fig. 2). DYRK1A itself is a multidomain protein containing a highly conserved kinase domain and a preceding DH domain that is characteristic of the Dyrk subfamily (Becker & Joost, 1999; Leder et al., 1999). In addition, DYRK1A also contains two different nuclear localization signals, a PEST region, a histidine-rich region and an extreme C-terminal region rich in serine and threonine (Galceran, de Graaf, Tejedor, & Becker, 2003) (Fig. 2A). The interest in understanding the biological function of this human serine threonine kinase is greater than ever due to its multifaceted role in a variety of cellular and physiological processes as shown in Fig. 1. Understanding the domain structure and intramolecular control, along with downstream protein effectors of DYRK1A, is of paramount importance as this will give us a clear basis for elucidating the complex cellular function and signaling pathways that this protein kinase participates in, in addition to clarifying how it is actually regulated (Fig. 2).

2. MINIBRAIN IN *DROSOPHILA*

Drosophila melanogaster (fruit flies) have been utilized as a model system for genetic research over 100 years, beginning with the discovery by Thomas Hunt Morgan that genes reside on chromosomes in 1910, leading to the Nobel Prize award for this work in 1933 (Morgan, 1910). Surprisingly

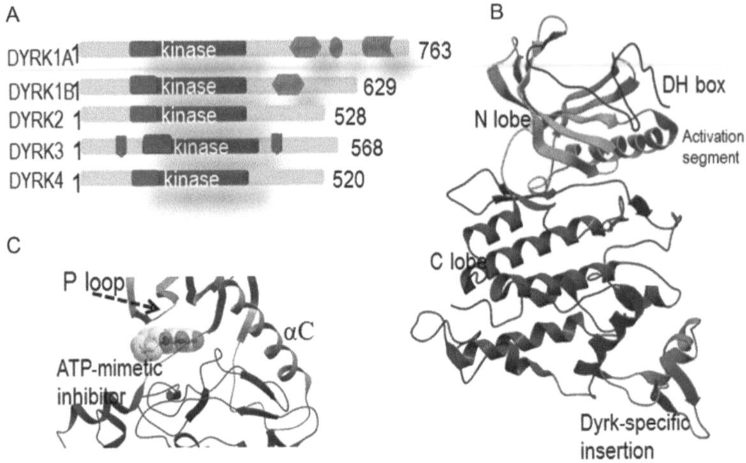

Fig. 2 The DYRK family and DYRK1A structure. (A) Domain architecture of DYRK kinases. DH domain characteristic to DYRK family is given using *red* (*gray* in the print version) *rectangles*. The PEST region is shown using *red* (*gray* in the print version) *hexagons*. *Red* (*gray* in the print version) *oval*, *cylinder*, and *pentagons* show histidine-rich region, serine/threonine-rich region, and nuclear localization sequences, respectively. (B) Overall structure of DYRK1A catalytic domain with N-terminal lobe shown in *orange* (*gray* in the print version). The C-terminal lobe is given in *green* (*dark gray* in the print version) while the DYRK-specific insert region is given in *purple* (*gray* in the print version). (C) A representative figure showing DYRK1A bound to ATP-mimetic small molecule inhibitor in the ATP-binding site in the hinge region of DYRK1A between N and C lobes. The P loop closing over the inhibitor is also shown in the figure along with activation segment αC helix.

for such a distantly related organism, more than 75% of known human disease genes were identified as having orthologs in flies (Reiter, Potocki, Chien, Gribskov, & Bier, 2001) and *Drosophila* is considered a useful model for studying the brain and neurological dysfunction. Indeed, the DYRK1A fly ortholog *minibrain* (*mnb*) was first discovered while screening for mutations implicated in structural brain defects, where it was shown to reduce central brain and optic lobe volume in adult flies, resulting in behavioral and locomotive defects (Fischbach & Heisenberg, 1984). Since this discovery, a plethora of reports regarding the *mnb* ortholog of DYRK1A have emerged and *mnb* mutant flies have proven exceedingly useful as model organisms in DS and other disorders, vastly improving knowledge on the actions of DYRK1A in health and disease. Later, the use of mammalian cells and murine mutants produced further reports on DYRK1A function in pathology, though initial mnb investigations were crucial, and *mnb* mutant flies remain routinely applied to research.

When the gene underlying the *mnb* mutation was characterized, it was found to code for a novel protein kinase with similarities to cyclin-dependent protein kinases involved in cell proliferation (Tejedor et al., 1995). Developmentally, *mnb* mRNA appeared to be expressed specifically in late larval tissue where new neurons are generated, with the *mnb* mutant displaying abnormal neuroblast organization in the outer proliferation center of the larval brain (Tejedor et al., 1995). These defects resulted in a decrease in central brain and optic lobe volume, in addition to abnormal leg and wing growth and altered behaviors (Tejedor et al., 1995). The *mnb* mutant flies exhibited poor freely walking visual pattern fixation and odor-discrimination learning, in addition to low negative geotaxis, walking activity, and speed. However, normal optomotor response in walking mode and in motion-induced landing response was observed (Fischbach & Heisenberg, 1984; Heisenberg, Borst, Wagner, & Byers, 1985; Tejedor et al., 1995).

In a screen for Dyrk1 substrates, *Drosophila* Mnb was also found to phosphorylate the substrates SNR1 and TRX, implicating roles for the Dyrk family in chromatin regulation (Kinstrie, Lochhead, Sibbet, Morrice, & Cleghon, 2006). In a different screen for genes regulating nuclear factor of activated T cells (NFAT) subcellular localization, Mnb was first identified in *Drosophila* as a novel regulator of NFAT by phosphorylating the conserved serine–proline repeat 3 (Gwack et al., 2006).

More recently, Mnb has been implicated in synaptic recycling (Chen et al., 2014). Mnb protein is localized in presynaptic terminals of the neuromuscular junction and appears to modulate synaptic terminal growth and size in a dose-dependent manner (Chen et al., 2014). Too, Mnb appears to be required for rapid synaptic vesicle recycling and this activity is likely to be mediated by synaptojanin (Synj) (Chen et al., 2014). Mnb phosphorylates Synj and in doing so, enhances Synj synaptic activity (Chen et al., 2014). Accordingly, Synj overexpression is able to mostly rescue *mnb* mutant synaptic recycling phenotypes (Chen et al., 2014). Similarly to many mammalian targets of Dyrk1A, *Drosophila* Synj is phosphorylated on multiple sites, including serine and threonine residues, though it lacks the typical consensus sequence for Dyrk1A/Mnb phosphorylation (Chen et al., 2014). *Drosophila* Synj, like its mammalian ortholog, does contain (S/T)P sites in the proline-rich domain (PRD) that could be the target for Mnb phosphorylation (Chen et al., 2014). Akin to Dyrk1A, Synj has been implicated in several neurological disorders including autism, PD, and DS.

RNAi ablation of *mnb* caused an increase in actin and long protrusions from central nervous system (CNS)-derived *Drosophila* cell bodies in culture, but not from cells derived from other tissues (Liu, Sims, & Baum, 2009). This supports recent evidence that Mnb phosphorylates Beta-tubulin and regulates microtubule dynamics in *Drosophila* neurons (Ori-McKenney et al., 2016). Indeed, microtubule polymerization is inhibited by Mnb phosphorylation of beta-tubulin at Serine-172 (Ori-McKenney et al., 2016). This regulation specifically organizes the morphology of actin-rich Class III dopaminergic neuron terminal dendrites during *Drosophila* larval brain development (Ori-McKenney et al., 2016); alteration of the dendritic architecture of these neurons by *mnb* mutation results in defects in their mechanosensory function (Ori-McKenney et al., 2016). This function appears conserved in mammals, with murine DYRK1A phosphorylating beta-tubulin at the S172 residue and inhibiting beta-tubulin polymerization in an ATP-dependent manner (Ori-McKenney et al., 2016).

Research addressing minibrain/DYRKIA function in *Drosophila* is ongoing. As shown in a recent study (Ori-McKenney et al., 2016), flies can be used to investigate the interacting genes and mechanisms of Mnb function in great detail within the nervous system, in numerous manners applicable to human disease. Further to this, *Drosophila* has been acknowledged as an excellent model organism for other areas of drug discovery and disease beyond the neurological sphere (Pandey & Nichols, 2011), including immune dysregulation (Dushay & Eldon, 1998; Kounatidis & Ligoxygakis, 2012).

3. ROLE OF DYRK1A IN INFECTION AND IMMUNITY

A range of host entities play antagonistic roles in dictating the propagation and replication of viruses in human beings (Rivas, Schmaling, & Gaglia, 2016; Sureau & Negro, 2016; Vincent, Ziehr, & Moorman, 2016). In human immunodeficiencies, macrophages have been shown to play a key role in disease control due to the crucial roles they play in rendering resting T cells permissive for viral infection (Kumar & Herbein, 2014).

A variety of host kinases have been investigated over the years (Bardou-Jacquet, Lorho, & Guyader, 2011; Kumar, Sharma, Ly, Parslow, & Liang, 2011; Martinez-Gil et al., 2013; Perwitasari, Yan, O'Donnell, Johnson, & Tripp, 2015) in hope of exploiting the possibility of validating them as therapeutic targets, since human protein kinases have proven to be the most successfully targeted class of molecules in the past decade (Muller, Chaikuad, Gray, & Knapp, 2015). Significant roles played by protein kinases such as Hck (Alvarado, Tarafdar, Yeh, & Smithgall, 2014; Cornall, Mak, Greenway, & Tachedjian, 2013; Wales et al., 2015), belonging to Src family kinases, and Bruton's tyrosine kinase (Guendel et al., 2015), as well as double-stranded RNA-dependent protein kinase (PKR) (Yoon et al., 2015) have been well established in HIV replication within macrophages. Other kinases including PRKACB, CDK5R2, and AK1 have also been implicated through a high-throughput gene knockdown study where over-expression of kinases upregulated HIV-1 long-term repeat (LTR) activity (Jiang, Zhang, Zhang, Liu, & Lu, 2014). The HIV-1 LTR is considered a critical promoter region for HIV-1 gene expression, with many host interactions occurring at this location.

Meanwhile, the HIV-1 Nef virulence factor interacts with multiple host cell-signaling proteins (Basmaciogullari & Pizzato, 2014; Percario, Ali, Mangino, & Affabris, 2015; Sugden, Bego, Pham, & Cohen, 2016; Verollet, Le Cabec, & Maridonneau-Parini, 2015). It is of interest to note Nef binds to the Src homology 3 domains of Src family kinases, which is thought to result in kinase activation (Alvarado et al., 2014; Saksela, Cheng, & Baltimore, 1995). The activation of this kinase is crucial to establish infectivity, replication, and MHC-I downregulation in HIV (Hung et al., 2007). Other Tec family kinases, including Itk, which has been shown to play significant role in replication and infection, are present in HIV target cells. Given the interest in understanding and targeting tyrosine kinases and ser/threonine kinases to combat HIV, it is interesting to note that DYRK1A has also recently been implicated in HIV-1 as a novel kinase target (Bol et al., 2011; Booiman, Loukachov, van Dort, van't Wout, & Kootstra, 2015; Kadri et al., 2015).

Though highly active antiretroviral treatment (HAART) has been one of the greatest achievements in tackling HIV/AIDS, host macrophages have still been shown to produce viruses under HAART conditions as a consequence of suboptimal antiviral tissue penetration, leading to reduced efficacy (Coiras, Lopez-Huertas, Perez-Olmeda, & Alcami, 2009). Efficient measures to circumvent this challenge are therefore consistently underway in the antiviral drug development community.

Intriguingly, a recent genome-wide association study has implicated DYRK1A in the replication of HIV-1 in macrophages. Based on SNP genotyping using monocyte-derived macrophages from about 400 post HIV-1 infection blood donors, this study clearly identified a strong association between in vitro HIV-1 replication in monocyte-derived macrophages and SNP rs12483205 in DYRK1A (Bol et al., 2011). A combined analysis of the initial and replication cohort increased the strength of the association ($p = 4.8461026$). Additionally, the study also showed this SNP to be associated with HIV-1 disease progression in vivo in two independent cohorts ($p=0.035$ and 0.0048). The above findings suggest that the kinase DYRK1A is involved in the replication of HIV-1, in vitro in macrophages as well as in vivo.

These observations are of great value considering that kinase inhibitors can be a real alternative to overcome the current problems in HIV-1 drug resistance and reduced efficacy. Indeed, a selective small molecule inhibitor of Itk kinase activity (BMS-509744) was shown to potently block wild-type HIV-1 infectivity and replication (Lin et al., 2004) and hence further therapeutic possibilities are being considered. A wealth of research into developing small molecule inhibitors against DYRK1A has resulted in testing and reporting a range of inhibitors with various chemical scaffolds and QSAR properties (Bharate, Yadav, & Vishwakarma, 2013; Pan, Wang, & Bryant, 2013). More promise on this avenue of investigation came from a follow-up study that focused on understanding the molecular mechanism through which DYRK1A regulates HIV-1 replication using primary PBMC and multiple cell lines (Booiman et al., 2015). It has been well established through seminal work conducted by Crabtree et al., the antagonistic role DYRK1A and DSCR1 play in NFAT neucleocytoplasmic shuttle and gene expression function (Arron et al., 2006). Again, in a study by Koostra et al., clear evidence is provided that DYRK1A inhibits HIV-1 LTR-driven transcription by limiting NFAT nuclear localization and consequently repressing downstream transcriptional activation (Booiman et al., 2015). A study using DYRK1A small molecule inhibitor INDY showed inhibition of DYRK1A resulted in NFAT shuttle into the nucleus, overall implicating inhibition and DYRK1A downregulation of DYRK1A in higher nuclear NFAT levels and increased binding of NFAT to the HIV-1 LTR (Booiman et al., 2015).

The novel role of DYRK1A in immune regulation is now becoming very apparent. An appropriate balanced differentiation of Th1 and Th17 cells from naïve CD4+ T helper cells, which is largely represented by differentiated regulatory T cells, is an important determinant of immune homeostasis. Dysregulation of this process has been implicated in various inflammatory diseases (Jager & Kuchroo, 2010). The recent work by Khor et al. using an unbiased chemical biology approach screening about 3000 chemical compounds, most of which have been approved for use in humans, has reported the identification of DYRK1A inhibitor harmine as a lead compound that clearly reduced the extent of experimentally induced inflammatory response (Khor et al., 2015). When DYRK1A was inhibited in mature T cells they tended to develop into antiinflammatory T_{reg} cells while impairing Th17 differentiation without affecting other known T_{reg}/Th17 differentiation pathways (Khor et al., 2015). It is therefore evident DYRK1A that represents a novel mechanistic node at the branch between commitment to either T_{reg} or Th17 lineages. This is of importance since this process is related to the progression of systemic autoimmunity and mucosal inflammation.

Neurodegenerative tauopathies have been reported to be associated with an aberrant splicing of Tau (Park, Ahn, & Gallo, 2016). The precise role of Tau aberrant splicing in HIV, however, has never been studied in a HIV setting. HIV-1 transactivator protein Tat is implicated in the neuronal damage that contributes to neurocognitive impairment affecting HIV/AIDS patients (Fields et al., 2015; Zucchini et al., 2013). A study by Kadri et al. utilizing patient samples, cultured cell lines and animal models found that Tat promotes TAU 3R upregulation through hyperphosphorylation of SC35 retained in nuclear speckles (Kadri et al., 2015). It is proposed that the cellular mechanism behind this involves Tat-mediated increased expression of DYRK1A. This study further indicates DYRK1A dysregulation as an early event in neuronal complications of HIV infection (Kadri et al., 2015). The study also clearly showed that HIV-1 Tat inhibits splicing factors that SC35-mediated Tau exon10 inclusion through up regulation of DYRK1A. This finding of DYRK1A's involvement in the neuropathology of AIDS patients is of great interest, since neuroinflammation and cognitive impairment caused by HIV-1 infection poses a great challenge to even those already undergoing retroviral treatment.

Furthermore, a pharmacological study investigating the molecular mechanisms induced by lactones and chalcones isolated from medicinal plants have clearly detected DYRK1A as one of the inhibited candidate molecules (Folmer et al., 2006). The mechanism underlying this inhibition was suggested to be through TNFα-induced activation of NFkB. This is certainly of importance since NFkB plays a very central role in orchestrating various immune and inflammatory pathways (Del Prete et al., 2011). Dysregulation of the NFkB-mediated cell survival and immunity has been established to be the hallmark of a variety of human inflammatory and infectious diseases (Del Prete et al., 2011; Folmer et al., 2006). It is therefore of great significance this nuclear kinase has now been implicated in signaling crosstalk with NFkB pathway.

4. CARDIOVASCULAR FUNCTION OF DYRK1A

Mechanisms underlying heart failure, including those caused by myocardial infarctions, have always attracted interest due to their associated clinical significance. Studies over the years have established the role of aberrant calcium signaling in the transmission of stress signaling cascades that mediate pathologic remodeling of the cardiovascular system (Frey, McKinsey, & Olson, 2000). Research focusing on the molecular mechanisms underlying cardiac remodeling have reported the important role of calcium/calmodulin-dependent protein kinase II isoforms as cellular regulators of this process (Colomer, Mao, Rockman, & Means, 2003; Hoch, Meyer, Hetzer, Krause, & Karczewski, 1999). It was subsequently discovered that alternative splicing factor (ASF), belonging to the highly conserved splicing regulating proteins called "Serine/Arginine-rich (SR) proteins," forms a component of alternative splicing programs regulated by a set of tissue-specific alternative splicing events during cardiac remodeling (Xu et al., 2005). DYRK1A is distributed in the nucleoplasm and localizes in nuclear speckles that serve as a storage/assembly site of SR proteins including ASF. DYRK1A indeed phosphorylates ASF and was previously shown to regulate alternative splicing of tau in DS. Expression analysis of DYRK1A revealed that this protein kinase is not only expressed in high levels in brain but also in heart (Okui et al., 1999). About 60% of children with DS have heart defects and among and there is a very high occurrence of congenital heart abnormalities (Stoll, Alembik, Dott, & Roth, 1998). Though this observation was reported a decade ago, the mechanism and involvement of DYRK1A in cardiac pathology remain poorly defined. Interestingly, a recent series of studies have endeavored to address this unsolved question. Initially, DYRK1A regulation of CaMKIIδ alternative splicing through ASF was shown to be involved in renovascular hypertension-induced cardiac hypertrophy (Lu, Zhu, Qin, Yu, & Sheng, 2012). A set of studies investigated whether Dyrk1A-ASF-CaMKIIδ-mediated signaling would play a key role in cardiac remodeling and possibly heart failure. The study used rat models to investigate this and determined enhanced expression of DYRK1A leads to ASF phosphorylation, consequently regulating CaMKIIδ. It is therefore evident that the Dyrk1A-ASF-CaMKIIδ signaling pathway is involved in the progression of heart failure following myocardial infarction. It is quite useful to note the small molecule DYRK1A inhibitors such as EGCG and harmine lead to inactivation of this Dyrk1A-ASF-CaMKIIδ signaling pathway, which provide cardiologists new avenues of treatment to consider for arresting the progression of myocardial infarction-induced heart failure (He, Yao, Sheng, & Zhu, 2015). In a follow-up study by Yao et al. to further investigate the role of Dyrk1A-ASF-CaMKIIδ signaling in Valsartan inhibition of cardiac hypertrophy in renovascular hypertensive rats (Yao, Qin, Zhu, & Sheng, 2016), it was deciphered that Valsartan inhibition of cardiac hypertrophy in renovascular hypertensive 2K1C rats was mediated, at least partly, by Dyrk1A-ASF-CaMKIIδ signaling (Yao et al., 2016).

It has been well established that the signaling circuit of calcineurin–NFAT transcription factor shuttling between nucleus and cytoplasm is controlled by the phosphorylation by DYRK1A and dephosphorylation by calcineurin, which is controlled by DSCR1 in a highly regulated circuit (Arron et al., 2006). NFAT signaling indeed plays a critical role not only in the immune (Xanthoudakis et al., 1996) and nervous systems (Furman et al., 2016; Vihma, Luhakooder, Pruunsild, & Timmusk, 2016; Wu, Ericson, Chao, & Low, 2010), but also in cardiovascular development (Zeidler et al., 2016) and pathological endothelial cell activation during angiogenesis or inflammation (Lighthouse & Small, 2016).

In cardiomyocytes, DYRK1A functions as a negative regulator of hypertrophy through NFAT phosphorylation (Kuhn et al., 2009) but its mechanistic function in the heart though this pathway remains poorly understood. Hille et al. designed a study to investigate the potential protective role of DYRK1A in cardiac hypertrophy in vivo using transgenic mice (Hille et al., 2016). The overexpression of DYRK1A caused severe dilated cardiomyopathy associated with congestive heart failure, leading to premature death. The study further identified that DYRK1A interacts with D-cyclin members and represses their protein levels in vitro and in vivo. Furthermore, forced expression of DYRK1A resulted in phosphorylation of Thr280 of Ccnd2 promoting proteosomal degradation. In cardiomyocytes, DYRK1A actually impaired cell cycle progression through suppression of Rb/E2f-signaling and reducing the expression of E2f target genes. The dysregulation of this pathway, with impaired cardiomyocyte proliferation, leads to cardiomyopathy and it was also reported that dose-specific DYRK1A expression and activity could be critical for hyperplastic and hypertrophic growth of the developing heart (Hille et al., 2016).

Cardiac troponins, including cardiac troponin T (cTnT), have been entities of interest for clinicians and researchers due to the role of the troponin isoforms as a biomarker for cardiac damage (Katus et al., 1989) but also as a marker of heart development (Drysdale, Tonissen, Patterson, Crawford, & Krieg, 1994; Sugimoto et al., 2015). cTnT isoform alternative splicing of exon 5 is developmentally regulated (Briggs & Schachat, 1993) and the exon5 containing troponin isoforms are expressed in the fetal and neonatal heart but are absent in adult heart (Parmacek & Leiden, 1989). The splicing factor SRp55 is involved in cTnT exon5 splicing is phosphorylated by DYRK1A (Lu & Yin, 2016). A recent study by Lu and Yin found DYRK1A interacted with SRp55 and enhanced its promotion of cTnT exon 5 inclusion. Furthermore, upon DYRK1A overexpression a shift from cTnT exon 5 inclusion to exclusion was found during development that was delayed in the heart of Ts65Dn mice. This study provides us the initial evidence for the role of DYRK1A in the neonatal cardiac development (Lu & Yin, 2016).

Cardiac hypertrophy has been shown to be a major risk factor for chronic heart failure in rat models of hyperhomocysteinemia (Joseph et al., 2003). As noted earlier, DYRK1A is a key regulator of the molecular pathway involved in cardiac hypertrophy. Investigation of DYRK1A expression in the heart of hyperhomocysteinemic rats found hyperhomocysteinemia actually induced ventricular cardiomyocyte hypertrophy while decreasing protein DYRK1A expression (Raaf et al., 2010). It was hypothesized that decreased DYRK1A expression could be consistent with decreased antihypertrophic effects of DYRK1A leading to cardiomyocyte hypertrophy in the case of hyperhomocysteinemia models. Finally, a study that focused on determining the role of microRNAs in myocardial disease onset and progression discovered miR-199b as a significant player in this process (da Costa Martins et al., 2010). MicroRNAs are a class of single stranded, noncoding RNAs of about 22 nucleotides in length. This study showed miR-199b is a direct calcineurin/NFAT target gene that increases during human heart failure and targets DYRK1A, constituting a pathogenic feedforward mechanism that affects calcineurin-responsive gene expression (da Costa Martins et al., 2010). This study clearly showed that mutant mice overexpressing miR-199b, or DYRK1A haploinsufficient mice, were sensitized to calcineurin/NFAT signaling, resulting in stress-induced cardiomegaly. This demonstrates that DYRK1A mediates cardiac cellular signaling and gene expression, and that inhibitors of DYRK1A in the form of miRNAs or small molecule inhibitors can be exploited as therapeutic options for treating heart failure. Through cell-signaling studies and animal models, these reports clearly indicate the varied role of DYRK1A in cardiac development, control of splicing, and cardiac myocyte hypertrophy.

5. ROLE OF DYRK1A IN CANCER

Although initial experimental evidence suggested key roles for DYRK1A in the development of DS symptoms (Guimera et al., 1996; Song et al., 1996), newer evidence has described DYRK1A as a potent inhibitor of apoptosis (Fujita et al., 2015; Guo, Williams, Schug, & Li, 2010; Laguna et al., 2008; Seifert, Allan, & Clarke, 2008; Seifert & Clarke, 2009) and as an important component of signaling pathways involved in tumor suppression (Fernandez-Martinez, Zahonero, & Sanchez-Gomez, 2015; Liu et al., 2014), with the capacity to mediate the activity of proteins involved in the progression of cancers (Di Vona et al., 2015; Liang, Chang, Wang, & Yu, 2008; Zhang et al., 2015). Intriguingly, the enhanced presence of DYRK1A in DS children is thought to increase risk of acute leukemias in children with DS, but to promote antitumorigenic effects in adult patients (Hasle, 2001). The mechanisms by which these contrasting effects operate remain obscure, though several models have been proposed (Birger & Izraeli, 2012; Liao, Zhou, Zhang, & Lu, 2012) and the emergence of newly identified DYRK1A substrates have enriched our understanding for the role of this kinase in a range of cellular environments. This section discusses the variety of mechanisms DYRK1A employs to promote cell proliferation and inhibit apoptotic pathways, with the aim of presenting our current knowledge of DYRK1A activity in cancer.

5.1 DYRK1A and the p53 Family

The p53 tumor suppressor protein plays essential roles in orchestrating vital cellular events such as cell cycle progression (Saavedra et al., 2015), apoptosis (Chipuk et al., 2004; Hagn et al., 2010; Jiang, Kon, et al., 2015; Leu, Dumont, Hafey, Murphy, & George, 2004; Matsuda et al., 2002; Mihara et al., 2003; Nakano & Vousden, 2001; Vaseva et al., 2012), autophagy (Baudot et al., 2016; Zhang, Xie, et al., 2016), and senescence (Anwar, Khosla, & Ramakrishna, 2016; Choi, Ryu, & Lim, 2016; Kim et al., 2016; Nicolai et al., 2015). Alongside its family members, p63 and p73, p53 regulates the expression of target genes at the level of transcription (Bourdon et al., 2005; Kern et al., 1991; Levrero et al., 2000) and is considered critically important in the prevention of neoplasia and tumor progression (Amelio & Melino, 2015; Haupt & Haupt, 2006; Hong, van den Heuvel, Prabhu, Zhang, & El-Deiry, 2014). Several independent reports now link the p53 family to DYRK1A, indicating that these proteins are involved in critically important signaling pathways known to cause cancer when dysregulated.

The first indication of a link between p53, DYRK1A and neoplasia arose in 2007, when DYRK1A overexpression was observed in HPV16 immortalized keratinocytes and malignant cervical lesions relative to primary foreskin keratinocytes and normal cervical tissue (Chang et al., 2007). Within HPV-derived cervical malignancies, genetic instability is thought to arise predominantly through the activity of the viral oncoprotein E6 on p53, leading to p53 degradation and ultimate p53 ineffectuality (Trave & Zanier, 2016; Werness, Levine, & Howley, 1990). In a 2007 study, ablation of overexpressed DYRK1A via siRNA-interference produced apoptosis in immortalized and malignant cells. However, a direct link between p53 and DYRK1A was not explicitly shown until 2011, when Zhang et al. demonstrated a p53-induced expression of miR-1246, a micro RNA, in response to DNA damage, producing a downstream reduction of DYRK1A in various mammalian cell lines (Zhang, Liao, Zeng, & Lu, 2011). Ablation of p53 increased DYRK1A levels, whereas an overexpression of miR-1246 mimics downregulated DYRK1A levels and prevented the nuclear import of GFP-tagged NFATc1, a substrate of DYRK1A involved in immune regulation, organ development, and implicated in tumorigenesis (Baek et al., 2009; Ryeom et al., 2008; Shaw et al., 1988; Wu et al., 2010; Xanthoudakis et al., 1996). This p53-mediated downregulation of DYRK1A and subsequent retention of NFATc1 was shown to correlate with an increased apoptotic rate in nonsmall cell lung cancer cells (Zhang et al., 2011).

The p53-induced expression of miR-1246 was later demonstrated to inhibit human hepatocellular carcinoma cells (Zhang et al., 2015), in part through direct inhibition of nuclear factor I/B (NFIB) by miR-1246 but possibly also through DYRK1A repression. Interestingly, miR-1246 was later shown to be upregulated in a dose-dependent manner to the presence of the soluble β-galactoside-binding animal lectin Galectin-9, resulting in downregulation of DYRK1A and a subsequent increase in apoptotic rate via caspase-9 (Fujita et al., 2015). However, it was not shown whether Galectin-9 bound directly to miR-1246. Thus, it is possible that the p53 and Galectin-9 modulation of DYRK1A through miR-1246 are linked. Indeed, the pathway involving miR-1246 and DYRK1 appears to be strongly evolutionarily conserved (Fujita et al., 2015).

Beyond p53, the suggestion exists of p73 regulation of DYRK1A and NFATc1 through miR-1246 or other means (Boominathan, 2010; Liao et al., 2012). This suggestion requires validation, though p73 exhibits strong structural and functional homology with p53 and it is thought that these family members may transcriptionally activate many of the same targets (Brandt, Petrovich, Joerger, & Veprintsev, 2009; Ciribilli et al., 2013; Dotsch, Bernassola, Coutandin, Candi, & Melino, 2010). Too, while p73 plays critical roles in brain and immune development (Belyi et al., 2010), DYRK1A is involved in neuron development (Becker & Sippl, 2011; Toiber et al., 2010) and NFATc1 acts as a crucial transcriptional regulator of numerous components of immune response, such as various cytokines and their receptors (Baek et al., 2009; Ryeom et al., 2008; Shaw et al., 1988; Wu et al., 2010; Xanthoudakis et al., 1996). A link between p73 and the DYRK1A–NFATc1 pathway would therefore logically explain the role of p73 in both brain and immune development. At present, however, this notion remains to be experimentally assessed.

5.2 DYRK1A in Apoptosis

The caspases are a family of endoproteases involved in the control of cellular inflammation and programmed cell death (apoptosis) (McIlwain, Berger, & Mak, 2013). Of this family, caspase-9 (Cas-9) is regarded as a critical initiator of the intrinsic (mitochondrial) pathway of apoptosis (Wurstle, Laussmann, & Rehm, 2012). This pathway is activated by a variety of cellular stresses, including DNA damage, accumulation of unfolded proteins, hypoxia, cytoskeletal disruption, and growth factor depravation (Brenner & Mak, 2009). However, developmental signals such as those derived from hormonal origin, can also induce this pathway of apoptosis (Brenner & Mak, 2009). Cas-9 generally exists in an inactive, procaspase monomer and requires activation by upstream signaling events to dimerize and become functionally active (Shiozaki, Chai, & Shi, 2002). Upon activation, Cas-9 is responsible for an apoptotic signaling cascade, processing and activating caspases 3 and 7 and ultimately resulting in the programmed demise of the cell (McIlwain et al., 2013). Apoptosis must be finely balanced. During normal development, correctly programmed cell death is a requisite for culling excessive cells and ensuring normal brain and limb development (Hakem et al., 1998; Kuida et al., 1998). However, aberrant upregulation of apoptosis can result in neurodegenerative disorders (Mattson, 2000; Roy et al., 1995). In contrast, a failure to activate apoptosis can lead to irregular cell growth and tumor progression. It is therefore critical that the signaling molecules involved in apoptosis are carefully orchestrated.

It is now known that DYRK1A interacts both directly and indirectly with Cas-9, acting as a potent inhibitor of Cas-9-mediated intrinsic apoptosis (Fujita et al., 2015; Laguna et al., 2008; Seifert et al., 2008; Seifert & Clarke, 2009). Previous studies have shown that inhibition of Cas-9 occurs by phosphorylation at threonine (T) 125 (Seifert et al., 2008), and that this phosphorylation is catalyzed by a number of kinases in a variety of contexts: by CDK1 during mitosis (Allan & Clarke, 2007), ERK1/2 in response to growth factors (Park et al., 2013) and p38 MAPK following hyperosmotic stress (Seifert & Clarke, 2009). However, DYRK1A has been demonstrated to both directly phosphorylate Cas-9T125 at a basal level and to sequester Cas-9 at the nucleus, further inhibiting its activity (Seifert et al., 2008). Depletion or inhibition of DYRK1A prevented Cas-9 phosphorylation and therefore its inhibition, indicating a critical role for DYRK1A in orchestrating basal Cas-9 inhibition (Laguna et al., 2008; Seifert et al., 2008).

DYRK1A also appears to inhibit apoptosis indirectly, via phosphorylation and subsequent activation of the NAD-dependent deacetylase sirtuin-1 (Sirt1) (Guo et al., 2010). Sirt1 links cellular metabolic status to gene regulation, directly connecting NAD+ hydrolysis to the deacetylation of histones and various transcription factors such as p53, FOXO, c-Myc, HIF-2a, E2F1, and NFkB (Sharma, Gautam, Costantini, Paladino, & Colonna, 2012). By direct phosphorylation of Sirt1 at threonine 522 (T522), DYRK1A leads to Sirt1-mediated deacetylation of p53, resulting in cell survival despite the presence of genotoxic stress. Ablation of DYRK1A restored activation of p53 and increased cell death in response to etoposide (Guo et al., 2010). Indeed, prior exhibition of DYRK1A overexpression in malignant cervical lesions and HPV16 immortalized keratinocytes revealed that DYRK1A overexpression results in suppression of the proapoptotic protein BAD and aberrant relocalization of the proapoptotic transcription factor forkhead in rhabdomyosarcoma (FKHR), thereby inhibiting FKHR activity (Chang et al., 2007).

In Drosophila, overexpression of minibrain (mnb) has been shown to promote organ growth in flies through inhibition of the Salvador–Warts–Hippo pathway, also known as the HIPPO pathway (Degoutin et al., 2013). This pathway is known to inhibit of proliferation and induce of apoptosis in flies and mammals (Badouel, Garg, & McNeill, 2009; Halder & Johnson, 2011; Harvey & Tapon, 2007), thus DYRK1A-mediated HIPPO inhibition is postulated to promote cell growth and inhibit apoptotic signaling. Indeed, as shown in Fig. 3, the DYRK1A fly ortholog minibrain has

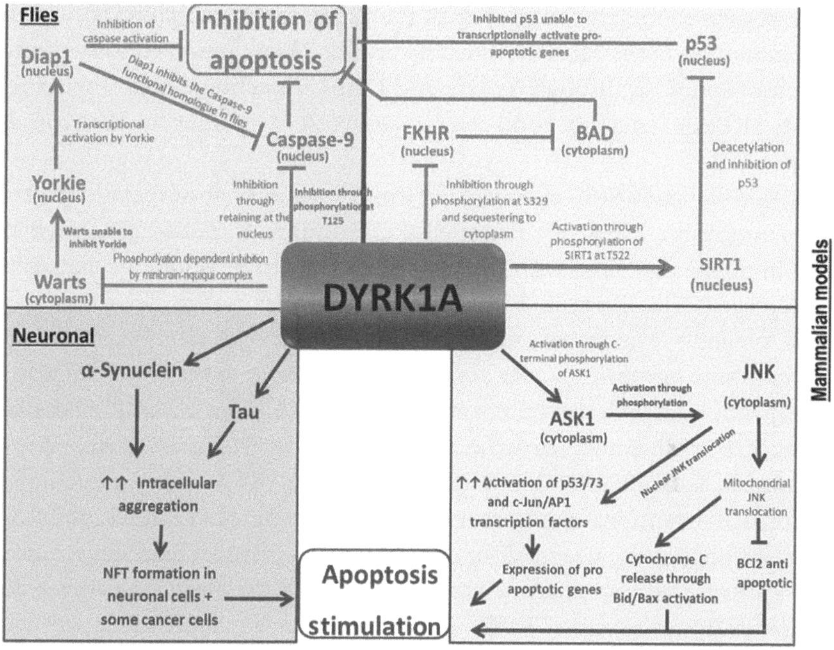

Fig. 3 Contrasting effects of DYRK1A on apoptotic signaling. The action of DYRK1A and its fly ortholog, minibrain, can influence both proapoptotic and antiapoptotic signaling. Here, *red* (*dark gray* in the print version) *lines* indicate actions leading to the inhibition of apoptosis through DYRK1A, while *green* (*black* in the print version) *lines* indicate actions ultimately leading to the stimulation of apoptosis. Through direct phosphorylation of caspase-9, DYRK1A is able to directly inhibit intrinsic apoptosis. Conversely, DYRK1A-mediated activation of SIRT1 leads to inhibition of p53 and subsequent downstream inhibition of the intrinsic pathway of apoptosis. Apoptosis is also inhibited by DYRK1A-mediated phosphorylation and inhibition of FKHR, leading to downstream inhibition of the proapoptotic protein BAD. In flies, the DYRK1A-ortholog minibrain forms a complex with riquiqui, leading to phosphorylation and inhibition of Warts. This renders Warts unable to inhibit Yorkie, which is able to transcriptionally activate the antiapoptotic caspase-interacting protein Diap1. Conversely, DYRK1A phosphorylation of ASK1 leads to downstream phosphorylation and activation of JNK in a proapoptotic pathway. JNK is able to translocate to the nucleus when activated, increasing the transcriptional activation of proapoptotic genes via its interaction several transcription factors, including p53/p73 and c-Jun/AP1. Activated JNK also translocates to the mitochondria, where it inhibits the antiapoptotic Bcl2 family and stimulates cytochrome *C* release through activation the Bid-Bax signaling. In neuronal cells and some cancer cells, DYRK1A is known to hyperphosphorylate the proteins alpha synuclein and Tau, increasing intracellular aggregation of these proteins and contributing to the formation of neurofibrillary tangles (NFTs). The presence of NFTs and increased intracellular aggregation is known to stimulate apoptosis.

been shown to form a complex with Riquiqui, leading to phosphorylation and inhibition of the HIPPO-signaling protein Warts and the ultimate inhibition of apoptosis through Yorkie and Diap1. Functional homologues for these HIPPO-signaling components are present in humans (Barron & Kagey, 2014).

Approximately 90% of cancer patients die as a result of metastatic progression (Spano, Heck, De Antonellis, Christofori, & Zollo, 2012), which is commonly associated with the loss of response to proapoptotic signals in these cells (Celia-Terrassa & Kang, 2016). Further to this, approximately 33% of solid tumors in adult patients appear inherently resistant to proapoptotic signaling and are consequently resistant to many conventional therapeutic approaches even prior to metastasis (Allen et al., 2011; Bruyere et al., 2011; Lefranc, Brotchi, & Kiss, 2005; Mathieu et al., 2009; Ristorcelli & Lombardo, 2010; Soengas & Lowe, 2003). Thus, the combination of conventional chemotherapy with therapeutic DYRK1A inhibitors has been previously suggested for the treatment of cancers exhibiting inherent resistance to proapoptotic signaling through the intrinsic pathway of apoptosis (Ionescu et al., 2012).

In contrast to this, however, several previous reports have suggested that DYRK1A can act as a proapoptotic signaling component in some contexts. Neuronal cells overexpressing DYRK1A display abnormal processing of several pathogenic proteins, resulting in intracellular aggregation of such proteins and ultimately an increase in apoptotic rate due to increased cellular stress (Park, Yang, Yoon, & Chung, 2007). Too, susceptibility to cell death is known to be increased in various cell types obtained from DS patients (Busciglio & Yankner, 1995; Roat et al., 2007).

Further to this, a 2011 report implicated DYRK1A as a proapoptotic protein able to regulate programmed cell death through facilitation of apoptosis signal-regulating kinase 1 (ASK1)-modulated signaling (Choi & Chung, 2011). This study reported the activation of DYRK1A by various toxic stimuli in human embryonic kidney (HEK293) cells, including hydrogen peroxide and TNFα. Subsequent activation of the ASK1-JNK signaling pathway in a DYRK1A dosage-dependent manner appeared to occur, with direct binding of DYRK1A to the C-terminal of ASK1 shown. This binding appeared to be enhanced by the presence of toxic stimuli in HEK293 cells. However, while the authors postulated that DYRK1A positively stimulates the JNK-ASK1 proapoptotic pathway, apoptotic assays were not performed and whether and how this DYRK1A stimulation truly results in programmed cell death remains unknown.

Taken together, these reports implicate DYRK1A as a dosage-dependent component of apoptotic signaling pathways, capable of either potently inhibiting apoptosis or, in some contexts, enhancing it.

5.3 DYRK1A as a Promoter of Cell Proliferation

Enhanced cell proliferation and survival are canonical features of neoplasia, contributing to cancer progression and metastasis (Spano et al., 2012). In recent years, DYRK1A has been demonstrated as a promoter of cell proliferation via several mechanisms. First, DYRK1A appears to positively regulate Wnt signaling in both *Xenopus laevis* embryos and mammalian cells (Hong et al., 2012). The Wnt signaling pathway is an evolutionarily conserved pathway involved in regulating critical aspects of cell migration, cell polarity, cell fate determination, and embryonic development (Inestrosa & Varela-Nallar, 2015; Kahn, 2014). Unsurprisingly, dysregulation of this pathway is associated with a variety of pathogenic states including cancer development. DYRK1A was shown to directly phosphorylate nuclear p120-catenin, resulting in repression of Kaiso activity and a subsequent dramatic increase in canonical Wnt target genes such as *wnt11* and *siamois* as detected by real-time PCR analysis. Use of classical in vivo assays (Park et al., 2006) indicated that DYRK1A overexpression also produced an increase in other canonical components of Wnt signaling, including ARVCF and δ-catenin. Taken together, these findings indicated a dose-dependent role for DYRK1A in positively regulating Wnt signaling through p120-catenin.

Further to influences on Wnt signaling, DYRK1A has been shown to regulate components of the Notch and Hedgehog signaling pathways. Both of these pathways are involved in the mediation of critical cellular processes such as cell proliferation, maintenance, survival, and differentiation. As such, dysregulation of both Notch and Hedgehog signaling has been linked to oncogenesis in a wide array of cell types. For instance, DYRK1A has been shown to increase the transcriptional activity of the Hedgehog (HH) signaling effector Gli1 (Morita, Lo Celso, Spencer-Dene, Zouboulis, & Watt, 2006; Shimokawa et al., 2013), a transcription factor first identified as overexpressed in some glioblastomas (Kinzler et al., 1987), and now recognized as a promoter of neoplasia in a variety of cell types when upregulated (Infante et al., 2015; Zhu & Lo, 2010). Similarly, DYRK1A has been shown to phosphorylate the HH signaling-related transcription factor STAT3 at Ser727 and increase its activity in COS7 cells (Wiechmann et al., 2003). STAT3 upregulation is exhibited in a range of malignancies, such to the extent that STAT3 inhibitors have been designed for chemotherapeutic means (Cafferkey & Chau, 2016; Peyser et al., 2016). Paradoxically, hallmark symptoms of DS are associated with reduced HH signaling activity (Roper et al., 2006). Furthermore, a recent study implicated DYRK1A as an inhibitor of endogenous HH signaling by negative regulation of ABLIM proteins, the actin cytoskeleton, and the transcriptional coactivator MKL1 (MAL) (Schneider et al., 2015). Thus, DYRK1A appears capable of activating or inhibiting several aspects of HH signaling. In contrast to this, DYRK1A has been shown to diminish notch signaling in neuronal cells via phosphorylation of Notch in the nuclear department, attenuating notch-mediated transcription (Fernandez-Martinez et al., 2009). Hyperactivity of the Notch pathway is typically considered oncogenic (Espinoza & Miele, 2013). Surprisingly, however, increased notch activity has been implicated in tumor suppression in myeloid malignancies (Ye et al., 2016). In light of this, DYRK1A phosphorylation and subsequent inhibition of Notch may discourage or encourage neoplasia, depending on cellular context and DYRK1A state.

A similar situation arises with the DYRK1A-interacting protein REST, which displays tumor suppressor activity in epithelial cancers but the capacity to promote neoplasia in neural tumors such as medulloblastoma (Taylor et al., 2012), neuroblastoma (Liang et al., 2014), and glioblastoma (Zhang, Li, et al., 2016). REST appears capable of transcriptionally activating DYRK1A by targeting a NRSE site in the human *DYRK1A* promoter region, and both REST and DYRK1A are coordinately expressed throughout neural development in the mouse brain (Kim, Jeong, et al., 2006). Further to this, dosage imbalance of DYRK1A has been shown to decrease REST expression via facilitation of the ubiquitination and subsequent proteasomal degradation of REST. Thus, it would seem that DYRK1A may function as a positive or negative mediator of REST, which could, in turn, contribute to tumor suppression or neoplasia.

In HPV16-infected cells, DYRK1A has been implicated in increasing cell survival and proliferation by phosphorylation of the HPV16 oncoprotein E7 (HPV16E7) at threonine 5 and threonine 7 (Liang et al., 2008). This phosphorylation was shown to increase the half-life of HPV16E7 and significantly increase infected cell colony formation. Too, HPV16E7 appeared to increase the transcriptional level of DYRK1A, ensuring enhanced stability of HPV16E7, and continuation of its prooncogenic effects. To continue the theme of viral oncogenesis, the infection of primary rodent cells with human adenovirus 5 (HAdV-5) has been shown to produce immortalized malignancies in a DYRK1A-regulatable manner. Here, DYRK1A appears to target the C-terminal region of the HAdV-5 oncogenic E1A protein, resulting in phosphorylation of E1A at serine 219 and subsequent attenuation of E1A signaling, leading to a diminishment of E1A's oncogenic activities (Glenewinkel et al., 2016). Thus, DYRK1A is able to act either to positively or negatively regulate oncogenic signaling in virally induced malignancies.

Perhaps most intriguingly, however, is the recent discovery that DYRK1A is recruited to the proximal promoter regions of growth-related genes in a manner that is independent of cell growth conditions (Di Vona et al., 2015). Here, DYRK1A functions as a RNA II polymerase kinase, targeting the C-terminal domain of RNA polymerase II for phosphorylation at serine 2 and serine 5. This phosphorylation was reduced upon shRNA-mediated depletion of DYRK1A. Thus, DYRK1A was found to specifically stimulate the transcription of several growth-related genes, many of which were reported to be implicated in ribosomal biogenesis and translational regulation. As DYRK1A has been previously implicated in cell cycle regulation through phosphorylation of several proteins such as cyclin D1 (Yabut, Domogauer, & D'Arcangelo, 2010), p27 (Soppa et al., 2014), or DREAM complex components (Litovchick, Florens, Swanson, Washburn, & DeCaprio, 2011), this new evidence enriches the role of DYRK1A as a potent modulator of both cell growth and division, able to influence several important signaling pathways and indeed, directly regulate several tumor suppressor or oncogenic proteins at the level of transcription.

5.4 DYRK1A in Myeloproliferative Disorders

Fascinatingly, DS patients exhibit an increased incidence of acute megakaryoblastic leukemia (AMLK) and ALL while simultaneously showing a markedly reduced risk of developing most solid tumors (Hasle, 2001). While there has been postulation that this may be related to the poor life expectancy of DS patients relative to the average age of solid tumor development (Liao et al., 2012), other reports indicate that an increased suppression of angiogenesis may lower solid tumor development in these patients. In endothelial cells, this suppression appears to occur via attenuation of VEGF–calcineurin–NFAT signaling in a DYRK1A–RCAN1 directed manner (Baek et al., 2009). Indeed, a reduced incidence of other angiogenesis-related diseases, such as atherosclerosis (Patterson, 2009) and diabetic retinopathy (Fulcher et al., 1998), occur among DS patients.

While DYRK1A inhibition of NFAT signaling is considered to promote tumor suppression in solid tumors, however, the inhibition of this same pathway through DYRK1A appears to deliver the opposite effect in some myeloproliferative disorders. NFAT is known to promote IL-2 during T-cell activation and NFAT overexpression has been observed in several hematological malignancies, with NFAT signaling considered to promote proliferation and survival in T-cell malignancies (Muller & Rao, 2010). Conversely, inhibition of NFAT signaling has been associated with megakaryopoiesis (Zaslavsky et al., 2013). Indeed, DYRK1A-mediated inhibition of NFAT factors produced an increased proliferation of neoplastic megakaryocytes in the Ts1Rhr trisomic murine model for DS (Malinge et al., 2012). In this model, DYRK1A shRNA-knockdown or inhibition by harmine resulted in a reduction in expanded megakaryocytes, suggesting a role for DYRK1A in stimulating AMLK cancer progression through mediation of NFAT signaling. However, in addition to targeting NFAT signaling, this murine model exhibited a link between DYRK1A and the transcription factor protein globin transcription factor 1 (GATA1) in megakaryocytic proliferation. Acquired mutations in GATA1 are commonly present in DS, resulting in the expression of shorter isoform of this protein (Roberts et al., 2013). While a strong link between GATA mutations and a predisposition toward leukemia is shown in DS patients, non-DS individuals harboring this mutation show no predisposition toward leukemias (Hollanda et al., 2006). GATA1 is normally considered an essential transcription factor in erythroid and megakaryocyte proliferation (Klusmann et al., 2010; Maroz et al., 2014). In the Ts1Rhr trisomic murine model, GATA1 appears to retain some of these functions but is unable to perform its normal repression of the oncogenic proteins Myc and E2F. E2F stimulates DYRK1A transcription (Maenz, Hekerman, Vela, Galceran, & Becker, 2008), thus a failure to inhibit E2F could drive the further expression of DYRK1A and accentuate the dosage-dependent effects of DYRK1A.

DYRK1A therefore appears to exhibit different, even antagonistic, functions depending on cellular context, DYRK1A dose and the availability and action of DYRK1A-interacting proteins. As such, DYRK1A can act as a double-edged sword: acting as a potent promoter of oncogenic activity in some contexts, while providing protective tumor suppressor properties in other situations.

6. UNDERSTANDING THE NEURONAL FUNCTION OF DYRK1A

Neurological roles for DYRK1A were indicated by initial reports revealing *DYRK1A* gene localization at the 21q22.2 region of chromosome 21.

Complete or partial trisomy of this chromosome region is directly related to the neuronal abnormalities of DS (Jiang, Liu, et al., 2015). DS has been established as the most common genetic cause of intellectual disability in the human population, with an occurrence rate of approximately 1 in 700 (Hook, 1981). There has been an exponential growth in investigating genes present in the Down syndrome critical region (DSCR) of chromosome 21 and its implication in neuropathies including dendrite atrophy, spine dysgenesis, cognitive disabilities, and tau-mediated neurodegeneration. Remarkably, small molecule inhibitors developed against DYRK1A kinase exhibited reductions or reversals of these neurodegenerative phenotypes when tested in cell lines and model organisms (Becker, Soppa, & Tejedor, 2014). Given the vast number of publications over the years reporting the role of DYRK1A in neuronal function and DS, this section of the review primarily focuses on the role of DYRK1A in tau-mediated and synuclein-mediated neuronal function. Recent reports implicating DYRK1A in Parkinson's disease (PD) will also be summarized.

6.1 DYRK1A in Neuronal Function

Two seminal studies from Crabtree et al. and Gwack et al. established a role for DYRK1A as a regulator of transcription factor NFAT's nucleocytoplasmic shuttle through phosphorylation and the antagonistic role DYRK1A plays with the DSCR1 gene (Arron et al., 2006; Gwack et al., 2006). The reports showed that the overexpression of DYRK1A and DSCR1 observed in trisomy 21 lead to destabilization of the regulatory circuit in the cell, causing dysregulated nuclear occupancy of NFAT. This is significant in light of the evidence for NFAT in controlling the expression of genes involved in musculoskeletal development, cardiac function, neuronal development, and immunity as shown through mouse models and genome-wide siRNA screening. Subsequent research that focused toward elucidating the roles of DYRK1A established the extreme gene dosage sensitivity of this protein during neural system development, where it controls proliferation, neurogenesis, neural differentiation, cell death, and synaptic plasticity.

Consequently, a variety of model systems have been used to illuminate the complex neuronal function controlled by DYRK1A.

6.2 DYRK1A in Tau-Related Neurodegenerative Diseases

A variety of human diseases involved with the loss of neuronal function have been linked to the microtubule protein Tau, which constitutes a major component of neurofibrillary tangles (NFTs) in many neurodegenerative disorders, including Alzheimer's disease (AD). Tau mutations implicated in frontotemporal dementia have resulted in targeted scientific studies endeavoring to understand the molecular mechanisms of aberrant tau in neurodegeneration (Grundke-Iqbal et al., 1986). As scientific evidence emerges on Tau-mediated neuronal death occurs even in the absence of NFTs, many studies now aim to investigate truncations, mutations, and posttranslational modifications of Tau as a source of Tau-mediated neurotoxicity. At the beginning of this decade, a high content siRNA kinome directed screening effort to determine the enzymes playing a significant role in AD-related Tau phosphorylation identified DYRK1A as a candidate gene involved in this process (Azorsa et al., 2010). This study first established DYRK1A's involvement in phosphorylating Tau at an AD-relevant epitope. Subsequently, several studies have established DYRK1A's crucial role in mediating neurodegeneration through direct phosphorylation of Tau and also through GSK3 kinase priming. It is indicated, therefore, that increased DYRK1A expression produces an additive effect on Tau phosphorylation that results in disruption of microtubule formation and Tau-mediated NFTs (see Fig. 4). This phenomenon of Tau phosphorylation by DYRK1A has wider implications, as related diseases include AD, frontal temporal

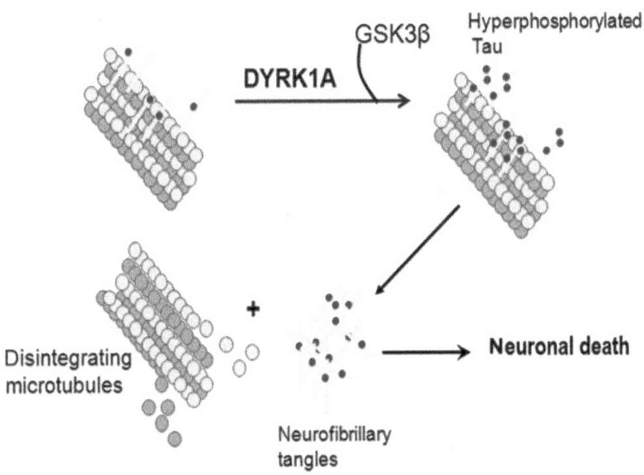

Fig. 4 Role of DYRK1A in tau hyperphosphorylation and neuronal diseases. In tauopathies, there is a reduction in the ability to bind microtubules and promote microtubule assembly. Hyperphosphorylated Tau by DYRK1A and DYRK1A-primed GSKβ contribute toward destabilized microtubule network, impaired axonal transport, results in neurofibrillary tangle (NFT) formation and eventually neuronal death. *Tau spheres* are given as *yellow* (*white* in the print version) *sticks* and the phosphorylation is shown using *small purple* (*dark gray* in the print version). Microtubule is shown using *cyan* (*white* in the print version) and *blue* (*dark gray* in the print version) *spheres.*

dementia with Parkinsonism linked to chromosome 17 (FTDP-17), progressive supranuclear palsy, Pick's disease, and corticobasal degeneration. Recent evidence has emerged for the presence of characteristic NFT patterns in different disease conditions. While NFTs are mainly distributed throughout the entorhinal region, AD patients exhibit significant depositions in the cortical brain and hippocampus. Conversely, FTDP-17 patients display NFTs in the frontal and temporal cortex and supranuclear palsy individuals suffer from depositions in the basal ganglia. Although the clinical presentations and NFT distributions vary, progressive accumulation of NFTs comprising hyperphosphorylated and insoluble Tau in a filamentous form, such as twisted or paired helical filaments, presents a common theme among these patients. It is noteworthy that aggregation of Tau into NFTs is directly related to neuronal death and a worsening neurodegenerative condition. Although convincing evidence from various studies has established the role of DYRK1A in AD, a case-controlled Spanish cohort of 634 is noteworthy. This study examined for genetic variations of DYRK1A by genotyping and haplotype tagging SNPs but detected no difference between the patient and control groups overall or after stratification by APOE epsilon4 allele. In contradiction to this, an unrelated study in the Japanese population reported a clear association between DYRK1A and AD risk. It is therefore necessary to make appropriate considerations and make further analysis to ascertain whether the DYRK1A effects are population based (Kimura et al., 2007; Vazquez-Higuera et al., 2009).

6.3 Tau Splicing Regulation by DYRK1A

DYRK1A plays a very interesting role in regulating Tau via controlling its splice variants. There are six different isoforms of Tau generated through alternative splicing that are expressed in the adult human brain. It is interesting to note that the Tau isoforms result from pre-mRNA of a single gene that contains 16 exons (Andreadis, Brown, & Kosik, 1992; Goedert, Spillantini, Jakes, Rutherford, & Crowther, 1989). Inclusion or exclusion of tau exon 10 (E10), that encodes the second microtubule-binding repeat, gives rise to tau isoforms with either four (4R) or three (3R) microtubule-binding repeats. Almost equal levels of 3R-Tau and 4R-Tau are expressed in the normal adult human brain. It was previously established that an imbalance between the 3R and 4R Tau can cause neurodegeneration and dementia (Goedert & Jakes, 2005; Liu & Gong, 2008; Wolfe, 2009). Indeed, several mutations in the *Tau* gene associated with FTDP-17 lead to dysregulation of tau E10 splicing. An altered 3R–4R ratio in the brain is also seen in corticobasal degeneration, Pick's disease and progressive supranuclear Palsy. As such, regulating the generation of these Tau isoforms and maintaining their concentration directly relates to neurodegeneration. DYRK1A plays a very significant role in multidimensional control of Tau isoforms through various splice factors. Alternative splicing is controlled by both exonic and intronic enhancers and silencers, which are controlled upstream by splice factors. In mammalian cells the splicing factors, which are protein regulators, fall into serine/arginine (SR)/SR-like and hnRNP proteins (Dreyfuss, Kim, & Kataoka, 2002; Graveley, 2000). It was initially demonstrated that DYRK1A phosphorylates ASF and inhibits its ability to promote Tau E10 inclusion (Shi et al., 2008). Interestingly, this work reported ASF phosphorylation on Ser-227, Ser-234, and Ser-238 residues by DYRK1A, driving the splice factor into nuclear speckles. By inhibition of association with the nascent transcripts, phosphorylation of ASF by DYRK1A results in Tau E10 exclusion leading to an unusual increase in 3R-tau levels, and causing an imbalance of 3R-4R tau isoforms in the DS brain. This dysregulation of tau E10 alternative splicing by DYRK1A represents an unusual NFT promotion mechanism, thereby offering a new avenue for therapeutic target development. Indeed, later work revealed that altered phosphorylation of ASF by overexpressed DYRK1A may contribute to the alternative splicing of the *Tau* gene and an increase by 2.68 times the 3R/4R ratio in DS/AD brains, thereby producing a several-fold increase in the number of 3R Tau-positive NFTs in DS/AD subjects compared with that in sporadic AD subjects (Wegiel et al., 2011). The data in this study supports the hypothesis that overexpressed DYRK1A in trisomy alters ASF phosphorylation could contribute to alternative splicing of E10, increased expression of 3R tau, and the early onset of neurofibrillary degeneration that has been reported in DS patients.

Another splice factor reported to be phosphorylated by DYRK1A is 9G8, which is a member of SR protein splice factors. The SR proteins possess a characteristic structure comprising of an N-terminal RNA recognition motif (RRM) and a C-terminal Arg-Ser-(RS)-rich domain. These groups of SR proteins have been shown to be tightly regulated by phosphorylation. DYRK1A interacted with and phosphorylated 9G8 in vitro and in live cells. A report by Ding et al. clearly shows DYRK1A regulation of 9G8 activity to modulate Tau E10 splicing. It was also shown E10 splicing was regulated by DYRK1A in a 9G8-independent manner in HEK293 cells (Ding et al., 2012).

SC35, another factor belonging to the spliceosome complex, is shown to be regulated by DYRK1A. Initial observations showed overexpression of DYRK1A changes the distribution of SC35 from speckles to a more diffuse pattern (Alvarez, Estivill, & de la Luna, 2003), suggesting that DYRK1A may also modulate SC35's function. Further investigation into this revealed SC35 was phosphorylated by DYRK1A and overexpression of DYRK1A suppressed SC35's ability to promote Tau E10 inclusion, while downregulation of DYRK1A promoted 4R-tau expression. Therefore, upregulation of DYRK1A in the DS or AD brain could cause a dysregulation of Tau E10 splicing via SC35, leading to the imbalance in 3R-tau and 4R-tau expression, thereby initiating or accelerating Tau pathology and causing neurofibrillary degeneration in the diseases (Qian et al., 2011).

Finally, there is recent evidence for DYRK1A regulating Tau E10 splicing through another splicing element. Another member of serine and arginine-rich proteins that play an important role in splicing of Tau exon 10 is serine/arginine rich (SRp55), which is also reported to be phosphorylated by DYRK1A. SRp55 is a SR protein with an approximate molecular mass of 55 kDa and reported to participate in constitutive and alternative mRNA splicing. SRp55 was originally shown to regulate the alternative splicing of epidermal growth factor, HIV-1, Tau, and Bim (Qian et al., 2013). It is now demonstrated that DYRK1A interacts with SRp55 through its N-terminal RRM domain and that DYRK1A is capable of phosphorylating SRp55 in the SRp55 proline-rich domain in cultured cells. SRp55 appeared to promote Tau exon 10 inclusion, which was mediated through its RS domain. Phosphorylation by DYRK1A inhibited SRp55 activity in promotion of Tau exon 10 inclusion. It can thus be postulated that overexpression of DYRK1A as seen in partial or complete chromosome trisomy would increase phosphorylation of SRp55, thereby inhibiting its Tau exon 10 splicing and resulting in 2R-tau overexpression and splice variant imbalance (Yin et al., 2012).

These studies indicate that DYRK1A functions as a master regulator of Tau exon 10 splicing through its phosphorylation capacity of various splice factors which could be specific to cellular and tissue requirements. Understanding the DYRK1A role in spliceosome function, particularly of tau exon 10, offers a new avenue for therapeutics, due to the direct correlation observed between Tau isoform imbalance and neurodegenerative function.

6.4 DYRK1A Is a Synuclein Regulator

Dementia with Lewy bodies (LB) is characterized by α-synuclein accumulation and degeneration of dopaminergic and cholinergic pathways. Alpha synuclein (α-synuclein) is the major component of LB and has been reported to occur in DS and AD patients. DYRK1A has been shown to directly phosphorylate α-synuclein and affects the intracellular α-synuclein inclusion formation when tested in hippocampal neuronal (H19-7) cells. Further examination revealed increased α-synuclein expression and phosphorylation, leading to increased intracytoplasmic aggregation. DYRK1A specifically phosphorylated α-synuclein on Ser-87 and the distinct hyperphosphorylated α-synuclein species was shown to be more neurotoxic when compared to unaltered α-synuclein. Considering that α-synuclein overexpression resulted in cell death when followed by basic fibroblast growth factor-induced neuronal differentiation, and that phosphorylated α-synuclein displays the above-mentioned phenotype, these initial findings taken together suggest that DYRK1A-mediated α-synuclein inclusion formation directly affects neuronal cell viability (Kim, Sung, et al., 2006).

Interestingly, α-synuclein is also an attractive candidate biomarker of PD: total and posttranslationally modified α-synuclein is strongly linked to the pathogenesis of sporadic and familial form of the disease (Polymeropoulos et al., 1997; Spillantini et al., 1997). Among several post-translational modifications to α-synuclein that occur in PD, phosphorylation at Ser^{129} [phosphorylated α-synuclein (PS-129)] is indicated to be critical in PD pathogenesis (Chen et al., 2009; Fujiwara et al., 2002). DYRK1A has been implicated to be one of the kinases that phosphorylate α-synuclein on Ser129. Increased concentration of α-synuclein in the brain, leading to aggregation and subsequently sequestration in LB, is a hallmark of PD. As such, continuous efforts are underway to identify novel mechanisms by which α-synuclein concentrations can be lowered. Although the total concentration of α-synuclein in cerebrospinal fluid (CSF) provides reasonable diagnostic sensitivity and specificity in differentiating PD patients from healthy controls, it is not sufficient to differentiate PD from other parkinsonian disorders including supranuclear palsy and multiple system atrophy. Moreover, α-synuclein levels do not correlate with the severity or progression of PD (Mollenhauer et al., 2011; Shi et al., 2011).

Originally α-synuclein-129 was demonstrated to play a toxic role in the α-synuclein transgenic Drosophila model, enhancing α-synuclein toxicity both in vivo and in vitro, possibly by increasing the formation of α-synuclein aggregates (Chen & Feany, 2005). This was later challenged when a study investigated the α-synuclein effect in rat transgenic models (Gorbatyuk et al., 2008). However, an investigation using oligodendroglial cells supported the critical role of S-129 phosphorylation in microtubule retraction mediated by α-synuclein (Kragh et al., 2009). This was investigated using inactive S129A and phosphomimetic S129D mutants. The cellular effect of expressing S129A equated to the loss of microtubule retraction, which is comparable with the role of Ser-129 phosphorylation in mediating α-synuclein inclusion formation and the unfolded protein response previously reported using SHSY5Y cells (Sugeno et al., 2008). However, the phosphomimetic mutant did not exhibit the same microtubule phenotype, the reasons for which remain obscure. Given the varied effect of S129 phosphorylation reported in other models and cell lines, the role of S129 phosphorylation in cellular control and its use in human disease pathogenesis and progression remained poorly defined.

Nevertheless, a translational study involving human CSF-based multicentric study determined that assessing both CSF pS-129 and total α-synuclein would enhance the accuracy of differential diagnosis among overlapping parkinsonian disorders. This will enable an improved diagnostic classification, especially at early disease stages, which is a great need in the current clinical scenario (Wang et al., 2012).

More recently, a pilot study designed to examine the association between DYRK1A- and α-synuclein-based dementias revealed fascinating insights. DNA samples from patients and controls with dementia were genotyped for the DYRK1A rs8126696 polymorphism. It was then determined rs8126696 polymorphism actually altered the risk of developing α-synuclein-associated dementias. This observation offers initial indications of DYRK1A as a potential therapeutic target for dementia, through the interaction of DYRK1A with several associated proteins (Jones, Aarsland, Londos, & Ballard, 2012).

6.5 Role of DYRK1A in Parkinson's Disease (PD)

PD is a neurodegenerative disorder of the CNS, typically presenting with depletion of dopaminergic neurones in the substantia nigra of the midbrain, resulting in progressive locomotor defects including shaking, rigidity, and difficulty walking. Cognitive and behavioral problems often arise as the disorder develops, including dementia and depression. Although it is believed that a range of dysregulated molecular mechanisms may contribute to PD several reports have implicated DYRK1A in the development of PD.

A recent epidemiological study of a cohort of 297 Chinese PD patients indicated that DYRK1A polymorphisms affect PD onset, with the *rs8126696 T* allele of DYRK1A associated with a significantly earlier age of PD onset (Fan et al., 2016). The associated molecular rationale behind this remains obscure, though prior manipulation of DYRK1A in mouse models has produced PD-like symptoms. For example, *Dyrk1A*$^{+/-}$ mice present with neuronal insufficiencies in specific adult brain regions (Benavides-Piccione et al., 2005; Dierssen et al., 2004). Notably, this includes the substantia nigra, which is canonically subject to deficient dopaminergic activity in PD. This haploinsufficient mouse model showed a normal generation of mesencephalic DA (mDA) neurones in the embryo, but reduction in these neurones postnatally, due to increased activation of the intrinsic pathway of apoptosis. Intriguingly, a mouse model carrying three copies of *DYRK1A* exhibited increased survival of mDA neurones at the same stages, even in the presence of MPTP intoxication. Thus, DYRK1A appears to influence the survival of mDA neurones in a dose-specific manner in regions quintessentially associated with PD.

It is likely that knowledge of DYRK1A signaling interactions in the context of PD is incomplete. However, using the kinase domain of DYRK1A as bait in a yeast two-hybrid system against an adult mouse brain cDNA library, Sitz et al. identified the SEPT4_v7 splice variant of the GTPase-binding protein SEPT4 as a DYRK1A-binding partner (Sitz et al., 2008). This interaction was confirmed by coimmunoprecipitation in HEK293 human embryonic kidney cells and subsequent immunofluorescent techiques exhibiting coexpression and colocalization of SEPT4 and DYRK1A in mouse neurones. Indeed, DYRK1A was found to phosphorylate SEPT4, an effect that was abolished with administration of the DYRK1A inhibitor harmine. These results implicated SEPT4 as DYRK1A substrate. This was considered significant in light of prior reports displaying attenuation of dopaminergic neurotransmission in mice lacking *Sept4* in addition to various indications that both SEPT4 and DYRK1A are involved in neuronal development and health. Strikingly, a recent report also demonstrated DYRK1A phosphorylation of the canonical PD-related protein parkin at serine 131 (Im & Chung, 2015). Colocalization of parkin and DYRK1A was observed, with DYRK1A-mediated phosphorylation of parkin shown to inhibit Parkin E3 ubiquitin ligase activity and attenuate its binding affinity for its ubiquitin-conjugating E2 enzyme and substrate. Moreover, the neuroprotective effects of parkin against 6-hydroxydopamine toxicity in dopaminergic SH-SY5Y cells were diminished in the presence of DYRK1A-induced parkin phosphorylation, suggesting that DYRK1A may act as a functional regulator of parkin, favoring PD pathogenesis under accommodating conditions.

7. CONCLUDING REMARKS

The range of substrates reported to be phosphorylated by DYRK1A include NFAT (Arron et al., 2006; Gwack et al., 2006), CREB (Yang, Ahn, & Chung, 2001), FKHR (Woods et al., 2001), Gli1 (Mao et al., 2002), STAT3 (Becker & Joost, 1999), SIRT (Guo et al., 2010), synuclein (Kim, Sung, et al., 2006), Tau (Qian et al., 2013), and dynamin (Huang et al., 2004) among others. From this range of target substrates and the wide functional consequences of DYRK1A interactions, it is therefore sensible to conclude that DYRK1A is capable of controlling a variety of molecular processes that underlie several physiological functions in both neuronal and nonneuronal signaling. Indeed, recent work analyzing DYRK1A's role in pancreatic function reported contrasting functions for this human enzyme. Some recent murine model studies reported that DYRK1A haploinsufficiency resulted in reduced pancreatic beta cell mass, severe glucose tolerance, and decreased beta cell populations (Shen et al., 2015). Conversely, other recent chemical inhibitor-based studies revealed very compelling evidence for DYRK1A's inhibition actually increasing human pancreatic beta cell proliferation (Wang et al., 2015).

As such, the overall molecular and regulatory controls exerted by DYRK1A in pancreatic biology remain to be completely established, alongside potential validation of DYRK1A as a diabetic target. It is therefore fair to conclude numerous evidences for crosstalk between DYRK1A-related signaling pathways in novel cellular processes is still emerging.

The wealth of novel small molecule inhibitors that have been successfully developed against DYRK1A authenticates this protein kinase as an attractive drug target in not only neurological disorders, but also several other medical conditions as discussed earlier. The availability of many high-resolution three-dimensional structures of DYRK1A in complex with various ATP-mimetic inhibitors (see Fig. 2C), as well as substrate peptide bound crystallographic data, provides an incredible resource for medicinal and synthetic chemists to continue with structure-based drug design and develop novel lead compounds that could be taken into clinical trials for future therapeutics in a range of DYRK1A-related pathologies.

REFERENCES

Allan, L. A., & Clarke, P. R. (2007). Phosphorylation of caspase-9 by CDK1/cyclin B1 protects mitotic cells against apoptosis. *Molecular Cell, 26*(2), 301–310.

Allen, T. D., Zhu, C. Q., Jones, K. D., Yanagawa, N., Tsao, M. S., & Bishop, J. M. (2011). Interaction between MYC and MCL1 in the genesis and outcome of non-small-cell lung cancer. *Cancer Research, 71*(6), 2212–2221.

Alvarado, J. J., Tarafdar, S., Yeh, J. I., & Smithgall, T. E. (2014). Interaction with the Src homology (SH3-SH2) region of the Src-family kinase Hck structures the HIV-1 Nef dimer for kinase activation and effector recruitment. *Journal of Biological Chemistry, 289*(41), 28539–28553.

Alvarez, M., Altafaj, X., Aranda, S., & de la Luna, S. (2007). DYRK1A autophosphorylation on serine residue 520 modulates its kinase activity via 14-3-3 binding. *Molecular Biology of the Cell, 18*(4), 1167–1178.

Alvarez, M., Estivill, X., & de la Luna, S. (2003). DYRK1A accumulates in splicing speckles through a novel targeting signal and induces speckle disassembly. *Journal of Cell Science, 116*(15), 3099–3107.

Amelio, I., & Melino, G. (2015). The p53 family and the hypoxia-inducible factors (HIFs): Determinants of cancer progression. *Trends in Biochemical Sciences, 40*(8), 425–434.

Andreadis, A., Brown, W. M., & Kosik, K. S. (1992). Structure and novel exons of the human tau gene. *Biochemistry, 31*(43), 10626–10633.

Anwar, T., Khosla, S., & Ramakrishna, G. (2016). Increased expression of SIRT2 is a novel marker of cellular senescence and is dependent on wild type p53 status. *Cell Cycle, 15*(14), 1–15.

Arron, J. R., Winslow, M. M., Polleri, A., Chang, C. P., Wu, H., Gao, X., et al. (2006). NFAT dysregulation by increased dosage of DSCR1 and DYRK1A on chromosome 21. *Nature, 441*(7093), 595–600.

Azorsa, D. O., Robeson, R. H., Frost, D., Meec hoovet, B., Brautigam, G. R., Dickey, C., et al. (2010). High-content siRNA screening of the kinome identifies kinases involved in Alzheimer's disease-related tau hyperphosphorylation. *BMC Genomics, 11*, 25.

Badouel, C., Garg, A., & McNeill, H. (2009). Herding Hippos: Regulating growth in flies and man. *Current Opinion in Cell Biology, 21*(6), 837–843.

Baek, K. H., Zaslavsky, A., Lynch, R. C., Britt, C., Okada, Y., Siarey, R. J., et al. (2009). Down's syndrome suppression of tumour growth and the role of the calcineurin inhibitor DSCR1. *Nature, 459*(7250), 1126–1130.

Bardou-Jacquet, E., Lorho, R., & Guyader, D. (2011). Kinase inhibitors in the treatment of chronic hepatitis C virus. *Gut, 60*(6), 879–880.

Barron, D. A., & Kagey, J. D. (2014). The role of the Hippo pathway in human disease and tumorigenesis. *Clinical and Translational Medicine, 3*, 25.

Basmaciogullari, S., & Pizzato, M. (2014). The activity of Nef on HIV-1 infectivity. *Frontiers in Microbiology, 5*, 232.

Baudot, A. D., Crighton, D., O'Prey, J., Somers, J., Sierra Gonzalez, P., & Ryan, K. M. (2016). p53 directly regulates the glycosidase FUCA1 to promote chemotherapy-induced cell death. *Cell Cycle.*

Becker, W., & Joost, H. G. (1999). Structural and functional characteristics of Dyrk, a novel subfamily of protein kinases with dual specificity. *Progress in Nucleic Acid Research and Molecular Biology, 62*, 1–17.

Becker, W., & Sippl, W. (2011). Activation, regulation, and inhibition of DYRK1A. *FEBS Journal, 278*(2), 246–256.

Becker, W., Soppa, U., & Tejedor, F. J. (2014). DYRK1A: A potential drug target for multiple Down syndrome neuropathologies. *CNS & Neurological Disorders Drug Targets, 13*(1), 26–33.

Becker, W., Weber, Y., Wetzel, K., Eirmbter, K., Tejedor, F. J., & Joost, H. G. (1998). Sequence characteristics, subcellular localization, and substrate specificity of DYRK-related kinases, a novel family of dual specificity protein kinases. *Journal of Biological Chemistry*, *273*(40), 25893–25902.

Belyi, V. A., Ak, P., Markert, E., Wang, H., Hu, W., Puzio-Kuter, A., et al. (2010). The origins and evolution of the p53 family of genes. *Cold Spring Harbor Perspectives in Biology*, *2*(6), a001198.

Benavides-Piccione, R., Dierssen, M., Ballesteros-Yanez, I., de Lagran, M. M., Arbones, M. L., Fotaki, V., et al. (2005). Alterations in the phenotype of neocortical pyramidal cells in the Dyrk1A+/− mouse. *Neurobiology of Disease*, *20*(1), 115–122.

Bharate, S. B., Yadav, R. R., & Vishwakarma, R. A. (2013). QSAR and pharmacophore study of Dyrk1A inhibitory meridianin analogs as potential agents for treatment of neurodegenerative diseases. *Medicinal Chemistry*, *9*(1), 152–161.

Birger, Y., & Izraeli, S. (2012). DYRK1A in Down syndrome: An oncogene or tumor suppressor? *Journal of Clinical Investigation*, *122*(3), 807–810.

Bol, S. M., Moerland, P. D., Limou, S., van Remmerden, Y., Coulonges, C., van Manen, D., et al. (2011). Genome-wide association study identifies single nucleotide polymorphism in DYRK1A associated with replication of HIV-1 in monocyte-derived macrophages. *PLoS One*, *6*(2), e17190.

Booiman, T., Loukachov, V. V., van Dort, K. A., van't Wout, A. B., & Kootstra, N. A. (2015). DYRK1A controls HIV-1 replication at a transcriptional level in an NFAT dependent manner. *PLoS One*, *10*(12), e0144229.

Boominathan, L. (2010). The tumor suppressors p53, p63, and p73 are regulators of micro-RNA processing complex. *PLoS One*, *5*(5), e10615.

Bourdon, J. C., Fernandes, K., Murray-Zmijewski, F., Liu, G., Diot, A., Xirodimas, D. P., et al. (2005). p53 isoforms can regulate p53 transcriptional activity. *Genes & Development*, *19*(18), 2122–2137.

Brandt, T., Petrovich, M., Joerger, A. C., & Veprintsev, D. B. (2009). Conservation of DNA-binding specificity and oligomerisation properties within the p53 family. *BMC Genomics*, *10*, 628.

Brenner, D., & Mak, T. W. (2009). Mitochondrial cell death effectors. *Current Opinion in Cell Biology*, *21*(6), 871–877.

Briggs, M. M., & Schachat, F. (1993). Origin of fetal troponin T: Developmentally regulated splicing of a new exon in the fast troponin T gene. *Developmental Biology*, *158*(2), 503–509.

Bruyere, C., Lonez, C., Duray, A., Cludts, S., Ruysschaert, J. M., Saussez, S., et al. (2011). Considering temozolomide as a novel potential treatment for esophageal cancer. *Cancer*, *117*(9), 2004–2016.

Busciglio, J., & Yankner, B. A. (1995). Apoptosis and increased generation of reactive oxygen species in Down's syndrome neurons in vitro. *Nature*, *378*(6559), 776–779.

Cafferkey, C., & Chau, I. (2016). Novel STAT 3 inhibitors for treating gastric cancer. *Expert Opinion on Investigational Drugs*, 1–9.

Celia-Terrassa, T., & Kang, Y. (2016). Distinctive properties of metastasis-initiating cells. *Genes & Development*, *30*(8), 892–908.

Chang, H. S., Lin, C. H., Yang, C. H., Yen, M. S., Lai, C. R., Chen, Y. R., et al. (2007). Increased expression of Dyrk1a in HPV16 immortalized keratinocytes enable evasion of apoptosis. *International Journal of Cancer*, *120*(11), 2377–2385.

Chen, C.-K., Bregere, C., Paluch, J., Lu, J. F., Dickman, D. K., & Chang, K. T. (2014). Activity-dependent facilitation of synaptojanin and synaptic vesicle recycling by the minibrain kinase. *Nature Communications*, *5*. Article number: 4246.

Chen, L., & Feany, M. B. (2005). Alpha-synuclein phosphorylation controls neurotoxicity and inclusion formation in a Drosophila model of Parkinson disease. *Nature Neuroscience*, *8*(5), 657–663.

Chen, L., Periquet, M., Wang, X., Negro, A., McLean, P. J., Hyman, B. T., et al. (2009). Tyrosine and serine phosphorylation of alpha-synuclein have opposing effects on neurotoxicity and soluble oligomer formation. *Journal of Clinical Investigation*, *119*(11), 3257–3265.

Chipuk, J. E., Kuwana, T., Bouchier-Hayes, L., Droin, N. M., Newmeyer, D., Schuler, M., et al. (2004). Direct activation of Bax by p53 mediates mitochondrial membrane permeabilization and apoptosis. *Science*, *303*(5660), 1010–1014.

Choi, H. K., & Chung, K. C. (2011). Dyrk1A positively stimulates ASK1-JNK signaling pathway during apoptotic cell death. *Experimental Neurobiology*, *20*(1), 35–44.

Choi, O. R., Ryu, M. S., & Lim, I. K. (2016). Shifting p53-induced senescence to cell death by TIS21/BTG2/Pc3 gene through posttranslational modification of p53 protein. *Cellular Signalling*, *28*(9), 1172–1185.

Ciribilli, Y., Monti, P., Bisio, A., Nguyen, H. T., Ethayathulla, A. S., Ramos, A., et al. (2013). Transactivation specificity is conserved among p53 family proteins and depends on a response element sequence code. *Nucleic Acids Research*, *41*(18), 8637–8653.

Coiras, M., Lopez-Huertas, M. R., Perez-Olmeda, M., & Alcami, J. (2009). Understanding HIV-1 latency provides clues for the eradication of long-term reservoirs. *Nature Reviews. Microbiology*, 7(11), 798–812.

Colomer, J. M., Mao, L., Rockman, H. A., & Means, A. R. (2003). Pressure overload selectively up-regulates Ca2+/calmodulin-dependent protein kinase II in vivo. *Molecular Endocrinology*, *17*(2), 183–192.

Cornall, A., Mak, J., Greenway, A., & Tachedjian, G. (2013). HIV-1 infection of T cells and macrophages are differentially modulated by virion-associated Hck: A Nef-dependent phenomenon. *Viruses*, *5*(9), 2235–2252.

da Costa Martins, P. A., Salic, K., Gladka, M. M., Armand, A. S., Leptidis, S., el Azzouzi, H., et al. (2010). MicroRNA-199b targets the nuclear kinase Dyrk1a in an auto-amplification loop promoting calcineurin/NFAT signalling. *Nature Cell Biology*, *12*(12), 1220–1227.

Degoutin, J. L., Milton, C. C., Yu, E., Tipping, M., Bosveld, F., Yang, L., et al. (2013). Riquiqui and minibrain are regulators of the hippo pathway downstream of Dachsous. *Nature Cell Biology*, *15*(10), 1176–1185.

Del Prete, A., Allavena, P., Santoro, G., Fumarulo, R., Corsi, M. M., & Mantovani, A. (2011). Molecular pathways in cancer-related inflammation. *Biochemia Medica (Zagreb)*, *21*(3), 264–275.

Di Vona, C., Bezdan, D., Islam, A. B., Salichs, E., Lopez-Bigas, N., Ossowski, S., et al. (2015). Chromatin-wide profiling of DYRK1A reveals a role as a gene-specific RNA polymerase II CTD kinase. *Molecular Cell*, *57*(3), 506–520.

Dierssen, M., Benavides-Piccione, R., Ballesteros-Yanez, I., de Lagran, M. M., Arbones, M. L., Fotaki, V., et al. (2004). Brain size and circuit complexity: A study of the cerebral cortex in DYRK1A mouse. *Behavior Genetics*, *34*(6), 637.

Ding, S., Shi, J., Qian, W., Iqbal, K., Grundke-Iqbal, I., Gong, C. X., et al. (2012). Regulation of alternative splicing of tau exon 10 by 9G8 and Dyrk1A. *Neurobiology of Aging*, *33*(7), 1389–1399.

Dotsch, V., Bernassola, F., Coutandin, D., Candi, E., & Melino, G. (2010). p63 and p73, the ancestors of p53. *Cold Spring Harbor Perspectives in Biology*, *2*(9), a004887.

Dreyfuss, G., Kim, V. N., & Kataoka, N. (2002). Messenger-RNA-binding proteins and the messages they carry. *Nature Reviews. Molecular Cell Biology*, *3*(3), 195–205.

Drysdale, T. A., Tonissen, K. F., Patterson, K. D., Crawford, M. J., & Krieg, P. A. (1994). Cardiac troponin I is a heart-specific marker in the Xenopus embryo: Expression during abnormal heart morphogenesis. *Developmental Biology*, *165*(2), 432–441.

Dushay, M. S., & Eldon, E. D. (1998). Drosophila immune responses as models for human immunity. *The American Journal of Human Genetics*, *62*(1), 10–14.

Espinoza, I., & Miele, L. (2013). Notch inhibitors for cancer treatment. *Pharmacology & Therapeutics*, *139*(2), 95–110.

Fan, K., Tang, B. S., Wang, Y. Q., Kang, J. F., Li, K., Liu, Z. H., et al. (2016). The GBA, DYRK1A and MS4A6A polymorphisms influence the age at onset of Chinese Parkinson patients. *Neuroscience Letters*, *621*, 133–136.

Fernandez-Martinez, J., Vela, E. M., Tora-Ponsioen, M., Ocana, O. H., Nieto, M. A., & Galceran, J. (2009). Attenuation of notch signalling by the Down-syndrome-associated kinase DYRK1A. *Journal of Cell Science*, *122*(Pt. 10), 1574–1583.

Fernandez-Martinez, P., Zahonero, C., & Sanchez-Gomez, P. (2015). DYRK1A: The double-edged kinase as a protagonist in cell growth and tumorigenesis. *Molecular & Cellular Oncology*, *2*(1), e970048.

Fields, J., Dumaop, W., Eleuteri, S., Campos, S., Serger, E., Trejo, M., et al. (2015). HIV-1 Tat alters neuronal autophagy by modulating autophagosome fusion to the lysosome: Implications for HIV-associated neurocognitive disorders. *Journal of Neuroscience*, *35*(5), 1921–1938.

Fischbach, K. F., & Heisenberg, M. (1984). Neurogenetics and behaviour in insects. *Journal of Experimental Biology*, *112*(1), 65–93.

Folmer, F., Blasius, R., Morceau, F., Tabudravu, J., Dicato, M., Jaspars, M., et al. (2006). Inhibition of TNFalpha-induced activation of nuclear factor kappaB by kava (Piper methysticum) derivatives. *Biochemical Pharmacology*, *71*(8), 1206–1218.

Frey, N., McKinsey, T. A., & Olson, E. N. (2000). Decoding calcium signals involved in cardiac growth and function. *Nature Medicine*, *6*(11), 1221–1227.

Fujita, K., Iwama, H., Sakamoto, T., Okura, R., Kobayashi, K., Takano, J., et al. (2015). Galectin-9 suppresses the growth of hepatocellular carcinoma via apoptosis in vitro and in vivo. *International Journal of Oncology*, *46*(6), 2419–2430.

Fujiwara, H., Hasegawa, M., Dohmae, N., Kawashima, A., Masliah, E., Goldberg, M. S., et al. (2002). alpha-Synuclein is phosphorylated in synucleinopathy lesions. *Nature Cell Biology*, *4*(2), 160–164.

Fulcher, T., Griffin, M., Crowley, S., Firth, R., Acheson, R., & O'Meara, N. (1998). Diabetic retinopathy in Down's syndrome. *British Journal of Ophthalmology*, *82*(4), 407–409.

Furman, J. L., Sompol, P., Kraner, S. D., Pleiss, M. M., Putman, E. J., Dunkerson, J., et al. (2016). Blockade of astrocytic calcineurin/NFAT signaling helps to normalize hippocampal synaptic function and plasticity in a rat model of traumatic brain injury. *Journal of Neuroscience*, *36*(5), 1502–1515.

Galceran, J., de Graaf, K., Tejedor, F. J., & Becker, W. (2003). The MNB/DYRK1A protein kinase: Genetic and biochemical properties. *Journal of Neural Transmission. Supplementum*, *67*, 139–148.

Glenewinkel, F., Cohen, M. J., King, C. R., Kaspar, S., Bamberg-Lemper, S., Mymryk, J. S., et al. (2016). The adaptor protein DCAF7 mediates the interaction of the adenovirus E1A oncoprotein with the protein kinases DYRK1A and HIPK2. *Scientific Reports*, *6*, 28241.

Goedert, M., & Jakes, R. (2005). Mutations causing neurodegenerative tauopathies. *Biochimica et Biophysica Acta*, *1739*(2–3), 240–250.

Goedert, M., Spillantini, M. G., Jakes, R., Rutherford, D., & Crowther, R. A. (1989). Multiple isoforms of human microtubule-associated protein tau: Sequences and localization in neurofibrillary tangles of Alzheimer's disease. *Neuron*, *3*(4), 519–526.

Gorbatyuk, O. S., Li, S., Sullivan, L. F., Chen, W., Kondrikova, G., Manfredsson, F. P., et al. (2008). The phosphorylation state of Ser-129 in human alpha-synuclein determines neurodegeneration in a rat model of Parkinson disease. *Proceedings of the National Academy of Sciences of the United States of America*, *105*(2), 763–768.

Graveley, B. R. (2000). Sorting out the complexity of SR protein functions. *RNA*, *6*(9), 1197–1211.

Grundke-Iqbal, I., Iqbal, K., Tung, Y. C., Quinlan, M., Wisniewski, H. M., & Binder, L. I. (1986). Abnormal phosphorylation of the microtubule-associated protein tau (tau) in Alzheimer cytoskeletal pathology. *Proceedings of the National Academy of Sciences of the United States of America*, *83*(13), 4913–4917.

Guendel, I., Iordanskiy, S., Sampey, G. C., Van Duyne, R., Calvert, V., Petricoin, E., et al. (2015). Role of Bruton's tyrosine kinase inhibitors in HIV-1-infected cells. *Journal of Neurovirology*, *21*(3), 257–275.

Guimera, J., Casas, C., Pucharcos, C., Solans, A., Domenech, A., Planas, A. M., et al. (1996). A human homologue of Drosophila minibrain (MNB) is expressed in the neuronal regions affected in Down syndrome and maps to the critical region. *Human Molecular Genetics*, *5*(9), 1305–1310.

Guo, X., Williams, J. G., Schug, T. T., & Li, X. (2010). DYRK1A and DYRK3 promote cell survival through phosphorylation and activation of SIRT1. *Journal of Biological Chemistry*, *285*(17), 13223–13232.

Gwack, Y., Sharma, S., Nardone, J., Tanasa, B., Iuga, A., Srikanth, S., et al. (2006). A genome-wide Drosophila RNAi screen identifies DYRK-family kinases as regulators of NFAT. *Nature*, *441*(7093), 646–650.

Hagn, F., Klein, C., Demmer, O., Marchenko, N., Vaseva, A., Moll, U. M., et al. (2010). BclxL changes conformation upon binding to wild-type but not mutant p53 DNA binding domain. *Journal of Biological Chemistry*, *285*(5), 3439–3450.

Hakem, R., Hakem, A., Duncan, G. S., Henderson, J. T., Woo, M., Soengas, M. S., et al. (1998). Differential requirement for caspase 9 in apoptotic pathways in vivo. *Cell*, *94*(3), 339–352.

Halder, G., & Johnson, R. L. (2011). Hippo signaling: Growth control and beyond. *Development*, *138*(1), 9–22.

Harvey, K., & Tapon, N. (2007). The Salvador–Warts–Hippo pathway—An emerging tumour-suppressor network. *Nature Reviews. Cancer*, 7(3), 182–191.

Hasle, H. (2001). Pattern of malignant disorders in individuals with Down's syndrome. *Lancet Oncology*, *2*(7), 429–436.

Haupt, S., & Haupt, Y. (2006). Importance of p53 for cancer onset and therapy. *Anti-Cancer Drugs*, *17*(7), 725–732.

He, J., Yao, J., Sheng, H., & Zhu, J. (2015). Involvement of the dual-specificity tyrosine phosphorylation-regulated kinase 1A-alternative splicing factor-calcium/calmodulin-dependent protein kinase IIdelta signaling pathway in myocardial infarction-induced heart failure of rats. *Journal of Cardiac Failure*, *21*(9), 751–760.

Heisenberg, M., Borst, A., Wagner, S., & Byers, D. (1985). Drosophila mushroom body mutants are deficient in olfactory learning. *Journal of Neurogenetics*, *2*(1), 1–30.

Hille, S., Dierck, F., Kuhl, C., Sosna, J., Adam-Klages, S., Adam, D., et al. (2016). Dyrk1a regulates the cardiomyocyte cell cycle via D-cyclin-dependent Rb/E2f-signalling. *Cardiovascular Research*, *110*(3), 381–394.

Himpel, S., Panzer, P., Eirmbter, K., Czajkowska, H., Sayed, M., Packman, L. C., et al. (2001). Identification of the autophosphorylation sites and characterization of their effects in the protein kinase DYRK1A. *Biochemical Journal*, *359*(Pt. 3), 497–505.

Hoch, B., Meyer, R., Hetzer, R., Krause, E. G., & Karczewski, P. (1999). Identification and expression of delta-isoforms of the multifunctional Ca2+/calmodulin-dependent

protein kinase in failing and nonfailing human myocardium. *Circulation Research*, *84*(6), 713–721.
Hollanda, L. M., Lima, C. S., Cunha, A. F., Albuquerque, D. M., Vassallo, J., Ozelo, M. C., et al. (2006). An inherited mutation leading to production of only the short isoform of GATA-1 is associated with impaired erythropoiesis. *Nature Genetics*, *38*(7), 807–812.
Hong, J. Y., Park, J. I., Lee, M., Munoz, W. A., Miller, R. K., Ji, H., et al. (2012). Down's-syndrome-related kinase Dyrk1A modulates the p120-catenin-Kaiso trajectory of the Wnt signaling pathway. *Journal of Cell Science*, *125*(Pt. 3), 561–569.
Hong, B., van den Heuvel, A. P., Prabhu, V. V., Zhang, S., & El-Deiry, W. S. (2014). Targeting tumor suppressor p53 for cancer therapy: Strategies, challenges and opportunities. *Current Drug Targets*, *15*(1), 80–89.
Hook, E. B. (1981). Rates of chromosome-abnormalities at different maternal ages. *Obstetrics and Gynecology*, *58*(3), 282–285.
Huang, Y., Chen-Hwang, M. C., Dolios, G., Murakami, N., Padovan, J. C., Wang, R., et al. (2004). Mnb/Dyrk1A phosphorylation regulates the interaction of dynamin 1 with SH3 domain-containing proteins. *Biochemistry*, *43*(31), 10173–10185.
Hung, C. H., Thomas, L., Ruby, C. E., Atkins, K. M., Morris, N. P., Knight, Z. A., et al. (2007). HIV-1 Nef assembles a Src family kinase-ZAP-70/Syk-PI3K cascade to down-regulate cell-surface MHC-I. *Cell Host & Microbe*, *1*(2), 121–133.
Im, E., & Chung, K. C. (2015). Dyrk1A phosphorylates parkin at Ser-131 and negatively regulates its ubiquitin E3 ligase activity. *Journal of Neurochemistry*, *134*(4), 756–768.
Inestrosa, N. C., & Varela-Nallar, L. (2015). Wnt signalling in neuronal differentiation and development. *Cell and Tissue Research*, *359*(1), 215–223.
Infante, P., Mori, M., Alfonsi, R., Ghirga, F., Aiello, F., Toscano, S., et al. (2015). Gli1/DNA interaction is a druggable target for Hedgehog-dependent tumors. *EMBO Journal*, *34*(2), 200–217.
Ionescu, A., Dufrasne, F., Gelbcke, M., Jabin, I., Kiss, R., & Lamoral-Theys, D. (2012). DYRK1A kinase inhibitors with emphasis on cancer. *Mini Reviews in Medicinal Chemistry*, *12*(13), 1315–1329.
Jager, A., & Kuchroo, V. K. (2010). Effector and regulatory T-cell subsets in autoimmunity and tissue inflammation. *Scandinavian Journal of Immunology*, *72*(3), 173–184.
Jiang, L., Kon, N., Li, T., Wang, S. J., Su, T., Hibshoosh, H., et al. (2015). Ferroptosis as a p53-mediated activity during tumour suppression. *Nature*, *520*(7545), 57–62.
Jiang, X., Liu, C., Yu, T., Zhang, L., Meng, K., Xing, Z., et al. (2015). Genetic dissection of the Down syndrome critical region. *Human Molecular Genetics*, *24*(22), 6540–6551.
Jiang, W. M., Zhang, X. Y., Zhang, Y. Z., Liu, L., & Lu, H. Z. (2014). A high throughput RNAi screen reveals determinants of HIV-1 activity in host kinases. *International Journal of Clinical and Experimental Pathology*, 7(5), 2229–2237.
Jones, E. L., Aarsland, D., Londos, E., & Ballard, C. (2012). A pilot study examining associations between DYRK1A and alpha-synuclein dementias. *Neurodegenerative Diseases*, *10*(1–4), 229–231.
Joseph, J., Joseph, L., Shekhawat, N. S., Devi, S., Wang, J., Melchert, R. B., et al. (2003). Hyperhomocysteinemia leads to pathological ventricular hypertrophy in normotensive rats. *American Journal of Physiology. Heart and Circulatory Physiology*, *285*(2), H679–H686.
Kadri, F., Pacifici, M., Wilk, A., Parker-Struckhoff, A., Del Valle, L., Hauser, K. F., et al. (2015). HIV-1-tat protein inhibits SC35-mediated tau exon 10 inclusion through up-regulation of DYRK1A kinase. *Journal of Biological Chemistry*, *290*(52), 30931–30946.
Kahn, M. (2014). Can we safely target the WNT pathway? *Nature Reviews. Drug Discovery*, *13*(7), 513–532.

Katus, H. A., Remppis, A., Looser, S., Hallermeier, K., Scheffold, T., & Kubler, W. (1989). Enzyme linked immuno assay of cardiac troponin T for the detection of acute myocardial infarction in patients. *Journal of Molecular and Cellular Cardiology*, *21*(12), 1349–1353.

Kern, S. E., Kinzler, K. W., Bruskin, A., Jarosz, D., Friedman, P., Prives, C., et al. (1991). Identification of P53 as a sequence-specific DNA-binding protein. *Science*, *252*(5013), 1708–1711.

Khor, B., Gagnon, J. D., Goel, G., Roche, M. I., Conway, K. L., Tran, K., et al. (2015). The kinase DYRK1A reciprocally regulates the differentiation of Th17 and regulatory T cells. *eLife*, *4*, e05920.

Kim, M. Y., Jeong, B. C., Lee, J. H., Kee, H. J., Kook, H., Kim, N. S., et al. (2006). A repressor complex, AP4 transcription factor and geminin, negatively regulates expression of target genes in nonneuronal cells. *Proceedings of the National Academy of Sciences of the United States of America*, *103*(35), 13074–13079.

Kim, J. E., Shin, J. S., Moon, J. H., Hong, S. W., Jung, D. J., Kim, J. H., et al. (2016). Foxp3 is a key downstream regulator of p53-mediated cellular senescence. *Oncogene*.

Kim, E. J., Sung, J. Y., Lee, H. J., Rhim, H., Hasegawa, M., Iwatsubo, T., et al. (2006). Dyrk1A phosphorylates alpha-synuclein and enhances intracellular inclusion formation. *Journal of Biological Chemistry*, *281*(44), 33250–33257.

Kimura, R., Kamino, K., Yamamoto, M., Nuripa, A., Kida, T., Kazui, H., et al. (2007). The DYRK1A gene, encoded in chromosome 21 Down syndrome critical region, bridges between beta-amyloid production and tau phosphorylation in Alzheimer disease. *Human Molecular Genetics*, *16*(1), 15–23.

Kinstrie, R., Lochhead, Pamela A., Sibbet, G., Morrice, N., & Cleghon, V. (2006). dDYRK2 and minibrain interact with the chromatin remodelling factors SNR1 and TRX. *Biochemical Journal*, *398*(Pt. 1), 45–54.

Kinzler, K. W., Bigner, S. H., Bigner, D. D., Trent, J. M., Law, M. L., O'Brien, S. J., et al. (1987). Identification of an amplified, highly expressed gene in a human glioma. *Science*, *236*(4797), 70–73.

Klusmann, J. H., Godinho, F. J., Heitmann, K., Maroz, A., Koch, M. L., Reinhardt, D., et al. (2010). Developmental stage-specific interplay of GATA1 and IGF signaling in fetal megakaryopoiesis and leukemogenesis. *Genes & Development*, *24*(15), 1659–1672.

Kounatidis, I., & Ligoxygakis, P. (2012). Drosophila as a model system to unravel the layers of innate immunity to infection. *Open Biology*, *2*(5), 120075.

Kragh, C. L., Lund, L. B., Febbraro, F., Hansen, H. D., Gai, W. P., El-Agnaf, O., et al. (2009). Alpha-synuclein aggregation and Ser-129 phosphorylation-dependent cell death in oligodendroglial cells. *Journal of Biological Chemistry*, *284*(15), 10211–10222.

Kuhn, C., Frank, D., Will, R., Jaschinski, C., Frauen, R., Katus, H. A., et al. (2009). DYRK1A is a novel negative regulator of cardiomyocyte hypertrophy. *Journal of Biological Chemistry*, *284*(25), 17320–17327.

Kuida, K., Haydar, T. F., Kuan, C. Y., Gu, Y., Taya, C., Karasuyama, H., et al. (1998). Reduced apoptosis and cytochrome c-mediated caspase activation in mice lacking caspase 9. *Cell*, *94*(3), 325–337.

Kumar, A., & Herbein, G. (2014). The macrophage: A therapeutic target in HIV-1 infection. *Molecular and Cellular Therapies*, *2*, 10.

Kumar, N., Sharma, N. R., Ly, H., Parslow, T. G., & Liang, Y. Y. (2011). Receptor tyrosine kinase inhibitors that block replication of influenza A and other viruses. *Antimicrobial Agents and Chemotherapy*, *55*(12), 5553–5559.

Laguna, A., Aranda, S., Barallobre, M. J., Barhoum, R., Fernandez, E., Fotaki, V., et al. (2008). The protein kinase DYRK1A regulates caspase-9-mediated apoptosis during retina development. *Developmental Cell*, *15*(6), 841–853.

Leder, S., Weber, Y., Altafaj, X., Estivill, X., Joost, H. G., & Becker, W. (1999). Cloning and characterization of DYRK1B, a novel member of the DYRK family of protein kinases. *Biochemical and Biophysical Research Communications, 254*(2), 474–479.

Lefranc, F., Brotchi, J., & Kiss, R. (2005). Possible future issues in the treatment of glioblastomas: Special emphasis on cell migration and the resistance of migrating glioblastoma cells to apoptosis. *Journal of Clinical Oncology, 23*(10), 2411–2422.

Leu, J. I. J., Dumont, P., Hafey, M., Murphy, M. E., & George, D. L. (2004). Mitochondrial p53 activates Bak and causes disruption of a Bak-Mcl1 complex. *Nature Cell Biology, 6*(5), 443–450.

Levrero, M., De Laurenzi, V., Costanzo, A., Sabatini, S., Gong, J., Wang, J. Y. J., et al. (2000). The p53/p63/p73 family of transcription factors: Overlapping and distinct functions. *Journal of Cell Science, 113*(10), 1661–1670.

Liang, Y. J., Chang, H. S., Wang, C. Y., & Yu, W. C. (2008). DYRK1A stabilizes HPV16E7 oncoprotein through phosphorylation of the threonine 5 and threonine 7 residues. *International Journal of Biochemistry & Cell Biology, 40*(11), 2431–2441.

Liang, J., Tong, P., Zhao, W., Li, Y., Zhang, L., Xia, Y., et al. (2014). The REST gene signature predicts drug sensitivity in neuroblastoma cell lines and is significantly associated with neuroblastoma tumor stage. *International Journal of Molecular Sciences, 15*(7), 11220–11233.

Liao, J. M., Zhou, X., Zhang, Y., & Lu, H. (2012). MiR-1246: A new link of the p53 family with cancer and Down syndrome. *Cell Cycle, 11*(14), 2624–2630.

Lighthouse, J. K., & Small, E. M. (2016). Transcriptional control of cardiac fibroblast plasticity. *Journal of Molecular and Cellular Cardiology, 91*, 52–60.

Lin, T. A., McIntyre, K. W., Das, J., Liu, C., O'Day, K. D., Penhallow, B., et al. (2004). Selective Itk inhibitors block T-cell activation and murine lung inflammation. *Biochemistry, 43*(34), 11056–11062.

Litovchick, L., Florens, L. A., Swanson, S. K., Washburn, M. P., & DeCaprio, J. A. (2011). DYRK1A protein kinase promotes quiescence and senescence through DREAM complex assembly. *Genes & Development, 25*(8), 801–813.

Liu, F., & Gong, C. X. (2008). Tau exon 10 alternative splicing and tauopathies. *Molecular Neurodegeneration, 3*, 8.

Liu, Q., Liu, N., Zang, S., Liu, H., Wang, P., Ji, C., et al. (2014). Tumor suppressor DYRK1A effects on proliferation and chemoresistance of AML cells by downregulating c-Myc. *PLoS One, 9*(6), e98853.

Liu, T., Sims, D., & Baum, B. (2009). Parallel RNAi screens across different cell lines identify generic and cell type-specific regulators of actin organization and cell morphology. *Genome Biology, 10*(3), 1–9.

Lochhead, P. A., Sibbet, G., Morrice, N., & Cleghon, V. (2005). Activation-loop autophosphorylation is mediated by a novel transitional intermediate form of DYRKs. *Cell, 121*(6), 925–936.

Lu, S., & Yin, X. (2016). Overexpression of Dyrk1A regulates cardiac troponin T splicing in cells and mice. *Biochemical and Biophysical Research Communications, 473*(4), 993–998.

Lu, J. Y., Zhu, J. H., Qin, X. T., Yu, X. H., & Sheng, H. Z. (2012). Role of valsartan on myocardial Calpain I, calcineurin and Ca/calmodulin-dependent protein kinase IIdelta expression of renovascular hypertensive rats. *Zhonghua Xin Xue Guan Bing Za Zhi, 40*(6), 511–515.

Maenz, B., Hekerman, P., Vela, E. M., Galceran, J., & Becker, W. (2008). Characterization of the human DYRK1A promoter and its regulation by the transcription factor E2F1. *BMC Molecular Biology, 9*, 30.

Malinge, S., Bliss-Moreau, M., Kirsammer, G., Diebold, L., Chlon, T., Gurbuxani, S., et al. (2012). Increased dosage of the chromosome 21 ortholog Dyrk1a promotes

megakaryoblastic leukemia in a murine model of Down syndrome. *Journal of Clinical Investigation, 122*(3), 948–962.
Mao, J., Maye, P., Kogerman, P., Tejedor, F. J., Toftgard, R., Xie, W., et al. (2002). Regulation of Gli1 transcriptional activity in the nucleus by Dyrk1. *Journal of Biological Chemistry, 277*(38), 35156–35161.
Maroz, A., Stachorski, L., Emmrich, S., Reinhardt, K., Xu, J., Shao, Z., et al. (2014). GATA1s induces hyperproliferation of eosinophil precursors in Down syndrome transient leukemia. *Leukemia, 28*(6), 1259–1270.
Martinez-Gil, L., Alamares-Sapuay, J. G., Ramana Reddy, M. V., Goff, P. H., Premkumar Reddy, E., & Palese, P. (2013). A small molecule multi-kinase inhibitor reduces influenza A virus replication by restricting viral RNA synthesis. *Antiviral Research, 100*(1), 29–37.
Mathieu, V., Pirker, C., Martin de Lassalle, E., Vernier, M., Mijatovic, T., DeNeve, N., et al. (2009). The sodium pump alpha1 sub-unit: A disease progression-related target for metastatic melanoma treatment. *Journal of Cellular and Molecular Medicine, 13*(9B), 3960–3972.
Matsuda, K., Yoshida, K., Taya, Y., Nakamura, K., Nakamura, Y., & Arakawa, H. (2002). p53AIP1 regulates the mitochondrial apoptotic pathway. *Cancer Research, 62*(10), 2883–2889.
Mattson, M. P. (2000). Apoptosis in neurodegenerative disorders. *Nature Reviews. Molecular Cell Biology, 1*(2), 120–129.
McIlwain, D. R., Berger, T., & Mak, T. W. (2013). Caspase functions in cell death and disease. *Cold Spring Harbor Perspectives in Biology, 5*(4), a008656.
Mihara, M., Erster, S., Zaika, A., Petrenko, O., Chittenden, T., Pancoska, P., et al. (2003). p53 has a direct apoptogenic role at the mitochondria. *Molecular Cell, 11*(3), 577–590.
Mollenhauer, B., Locascio, J. J., Schulz-Schaeffer, W., Sixel-Doring, F., Trenkwalder, C., & Schlossmacher, M. G. (2011). alpha-Synuclein and tau concentrations in cerebrospinal fluid of patients presenting with parkinsonism: A cohort study. *Lancet Neurology, 10*(3), 230–240.
Morgan, T. H. (1910). Sex limited inheritance in Drosophila. *Science, 32*(812), 120–122.
Morita, K., Lo Celso, C., Spencer-Dene, B., Zouboulis, C. C., & Watt, F. M. (2006). HAN11 binds mDia1 and controls GLI1 transcriptional activity. *Journal of Dermatological Science, 44*(1), 11–20.
Muller, S., Chaikuad, A., Gray, N. S., & Knapp, S. (2015). The ins and outs of selective kinase inhibitor development. *Nature Chemical Biology, 11*(11), 818–821.
Muller, M. R., & Rao, A. (2010). NFAT, immunity and cancer: A transcription factor comes of age. *Nature Reviews. Immunology, 10*(9), 645–656.
Nakano, K., & Vousden, K. H. (2001). PUMA, a novel proapoptotic gene, is induced by p53. *Molecular Cell, 7*(3), 683–694.
Nicolai, S., Rossi, A., Di Daniele, N., Melino, G., Annicchiarico-Petruzzelli, M., & Raschella, G. (2015). DNA repair and aging: The impact of the p53 family. *Aging (Albany NY), 7*(12), 1050–1065.
Okui, M., Ide, T., Morita, K., Funakoshi, E., Ito, F., Ogita, K., et al. (1999). High-level expression of the Mnb/Dyrk1A gene in brain and heart during rat early development. *Genomics, 62*(2), 165–171.
Ori-McKenney, K. M., McKenney, R. J., Huang, H. H., Li, T., Meltzer, S., Jan, L. Y., et al. (2016). Phosphorylation of beta-tubulin by the Down syndrome kinase, minibrain/DYRK1a, regulates microtubule dynamics and dendrite morphogenesis. *Neuron, 90*(3), 551–563.
Pan, Y., Wang, Y., & Bryant, S. H. (2013). Pharmacophore and 3D-QSAR characterization of 6-arylquinazolin-4-amines as Cdc2-like kinase 4 (Clk4) and dual specificity tyrosine-phosphorylation-regulated kinase 1A (Dyrk1A) inhibitors. *Journal of Chemical Information and Modeling, 53*(4), 938–947.

Pandey, U. B., & Nichols, C. D. (2011). Human disease models in Drosophila melanogaster and the role of the fly in therapeutic drug discovery. *Pharmacological Reviews*, *63*(2), 411–436.

Park, S. A., Ahn, S. I., & Gallo, J. M. (2016). Tau mis-splicing in the pathogenesis of neurodegenerative disorders. *BMB Reports*. Epub ahead of print.

Park, G. B., Choi, Y., Kim, Y. S., Lee, H. K., Kim, D., & Hur, D. Y. (2013). ROS and ERK1/2-mediated caspase-9 activation increases XAF1 expression in dexamethasone-induced apoptosis of EBV-transformed B cells. *International Journal of Oncology*, *43*(1), 29–38.

Park, J. I., Ji, H., Jun, S., Gu, D., Hikasa, H., Li, L., et al. (2006). Frodo links dishevelled to the p120-catenin/Kaiso pathway: Distinct catenin subfamilies promote Wnt signals. *Developmental Cell*, *11*(5), 683–695.

Park, J., Yang, E. J., Yoon, J. H., & Chung, K. C. (2007). Dyrk1A overexpression in immortalized hippocampal cells produces the neuropathological features of Down syndrome. *Molecular and Cellular Neurosciences*, *36*(2), 270–279.

Parmacek, M. S., & Leiden, J. M. (1989). Structure and expression of the murine slow/cardiac troponin C gene. *Journal of Biological Chemistry*, *264*(22), 13217–13225.

Patterson, D. (2009). Molecular genetic analysis of Down syndrome. *Human Genetics*, *126*(1), 195–214.

Percario, Z. A., Ali, M., Mangino, G., & Affabris, E. (2015). Nef, the shuttling molecular adaptor of HIV, influences the cytokine network. *Cytokine & Growth Factor Reviews*, *26*(2), 159–173.

Perwitasari, O., Yan, X. Z., O'Donnell, J., Johnson, S., & Tripp, R. A. (2015). Repurposing kinase inhibitors as antiviral agents to control influenza A virus replication. *Assay and Drug Development Technologies*, *13*(10), 638–649.

Peyser, N. D., Wang, L., Zeng, Y., Acquafondata, M., Freilino, M. L., Li, H., et al. (2016). STAT3 as a chemoprevention target in carcinogen-induced head and neck squamous cell carcinoma. *Cancer Prevention Research (Philadelphia, PA)*. http://www.ncbi.nlm.nih.gov/pubmed/27267892.

Polymeropoulos, M. H., Lavedan, C., Leroy, E., Ide, S. E., Dehejia, A., Dutra, A., et al. (1997). Mutation in the alpha-synuclein gene identified in families with Parkinson's disease. *Science*, *276*(5321), 2045–2047.

Qian, W., Jin, N., Shi, J., Yin, X., Jin, X., Wang, S., et al. (2013). Dual-specificity tyrosine phosphorylation-regulated kinase 1A (Dyrk1A) enhances tau expression. *Journal of Alzheimer's Disease*, *37*(3), 529–538.

Qian, W., Liang, H., Shi, J., Jin, N., Grundke-Iqbal, I., Iqbal, K., et al. (2011). Regulation of the alternative splicing of tau exon 10 by SC35 and Dyrk1A. *Nucleic Acids Research*, *39*(14), 6161–6171.

Raaf, L., Noll, C., Cherifi, M., Benazzoug, Y., Delabar, J. M., & Janel, N. (2010). Hyperhomocysteinemia-induced Dyrk1a downregulation results in cardiomyocyte hypertrophy in rats. *International Journal of Cardiology*, *145*(2), 306–307.

Reiter, L. T., Potocki, L., Chien, S., Gribskov, M., & Bier, E. (2001). A systematic analysis of human disease-associated gene sequences in Drosophila melanogaster. *Genome Research*, *11*(6), 1114–1125.

Ristorcelli, E., & Lombardo, D. (2010). Targeting notch signaling in pancreatic cancer. *Expert Opinion on Therapeutic Targets*, *14*(5), 541–552.

Rivas, H. G., Schmaling, S. K., & Gaglia, M. M. (2016). Shutoff of host gene expression in influenza A virus and herpesviruses: Similar mechanisms and common themes. *Viruses*, *8*(4), e102.

Roat, E., Prada, N., Ferraresi, R., Giovenzana, C., Nasi, M., Troiano, L., et al. (2007). Mitochondrial alterations and tendency to apoptosis in peripheral blood cells from children with Down syndrome. *FEBS Letters*, *581*(3), 521–525.

Roberts, I., Alford, K., Hall, G., Juban, G., Richmond, H., Norton, A., et al. (2013). GATA1-mutant clones are frequent and often unsuspected in babies with Down syndrome: Identification of a population at risk of leukemia. *Blood*, *122*(24), 3908–3917.

Roper, R. J., Baxter, L. L., Saran, N. G., Klinedinst, D. K., Beachy, P. A., & Reeves, R. H. (2006). Defective cerebellar response to mitogenic Hedgehog signaling in Down [corrected] syndrome mice. *Proceedings of the National Academy of Sciences of the United States of America*, *103*(5), 1452–1456.

Roy, N., Mahadevan, M. S., Mclean, M., Shutler, G., Yaraghi, Z., Farahani, R., et al. (1995). The gene for neuronal apoptosis inhibitory protein is partially deleted in individuals with spinal muscular-atrophy. *Cell*, *80*(1), 167–178.

Ryeom, S., Baek, K. H., Rioth, M. J., Lynch, R. C., Zaslavsky, A., Birsner, A., et al. (2008). Targeted deletion of the calcineurin inhibitor DSCR1 suppresses tumor growth. *Cancer Cell*, *13*(5), 420–431.

Saavedra, K., Valbuena, J., Olivares, W., Marchant, M. J., Rodriguez, A., Torres-Estay, V., et al. (2015). Loss of expression of reprimo, a p53-induced cell cycle arrest gene, correlates with invasive stage of tumor progression and p73 expression in gastric cancer. *PLoS One*, *10*(5), e0125834.

Saksela, K., Cheng, G., & Baltimore, D. (1995). Proline-rich (PxxP) motifs in HIV-1 Nef bind to SH3 domains of a subset of Src kinases and are required for the enhanced growth of Nef+ viruses but not for down-regulation of CD4. *EMBO Journal*, *14*(3), 484–491.

Schneider, P., Bayo-Fina, J. M., Singh, R., Kumar Dhanyamraju, P., Holz, P., Baier, A., et al. (2015). Identification of a novel actin-dependent signal transducing module allows for the targeted degradation of GLI1. *Nature Communications*, *6*, 8023.

Seifert, A., Allan, L. A., & Clarke, P. R. (2008). DYRK1A phosphorylates caspase 9 at an inhibitory site and is potently inhibited in human cells by harmine. *FEBS Journal*, *275*(24), 6268–6280.

Seifert, A., & Clarke, P. R. (2009). p38alpha- and DYRK1A-dependent phosphorylation of caspase-9 at an inhibitory site in response to hyperosmotic stress. *Cellular Signalling*, *21*(11), 1626–1633.

Sharma, A., Gautam, V., Costantini, S., Paladino, A., & Colonna, G. (2012). Interactomic and pharmacological insights on human sirt-1. *Frontiers in Pharmacology*, *3*, 40.

Shaw, J. P., Utz, P. J., Durand, D. B., Toole, J. J., Emmel, E. A., & Crabtree, G. R. (1988). Identification of a putative regulator of early T cell activation genes. *Science*, *241*(4862), 202–205.

Shen, W., Taylor, B., Jin, Q., Nguyen-Tran, V., Meeusen, S., Zhang, Y. Q., et al. (2015). Inhibition of DYRK1A and GSK3B induces human beta-cell proliferation. *Nature Communications*, *6*, 8372.

Shi, M., Bradner, J., Hancock, A. M., Chung, K. A., Quinn, J. F., Peskind, E. R., et al. (2011). Cerebrospinal fluid biomarkers for Parkinson disease diagnosis and progression. *Annals of Neurology*, *69*(3), 570–580.

Shi, J., Zhang, T., Zhou, C., Chohan, M. O., Gu, X., Wegiel, J., et al. (2008). Increased dosage of Dyrk1A alters alternative splicing factor (ASF)-regulated alternative splicing of tau in Down syndrome. *Journal of Biological Chemistry*, *283*(42), 28660–28669.

Shimokawa, T., Rahman, M. F., Tostar, U., Sonkoly, E., Stahle, M., Pivarcsi, A., et al. (2013). RNA editing of the GLI1 transcription factor modulates the output of Hedgehog signaling. *RNA Biology*, *10*(2), 321–333.

Shiozaki, E. N., Chai, J., & Shi, Y. (2002). Oligomerization and activation of caspase-9, induced by Apaf-1 CARD. *Proceedings of the National Academy of Sciences of the United States of America*, *99*(7), 4197–4202.

Sitz, J. H., Baumgartel, K., Hammerle, B., Papadopoulos, C., Hekerman, P., Tejedor, F. J., et al. (2008). The Down syndrome candidate dual-specificity tyrosine phosphorylation-regulated kinase 1A phosphorylates the neurodegeneration-related septin 4. *Neuroscience*, *157*(3), 596–605.

Soengas, M. S., & Lowe, S. W. (2003). Apoptosis and melanoma chemoresistance. *Oncogene*, *22*(20), 3138–3151.

Song, W. J., Sternberg, L. R., Kasten-Sportes, C., Keuren, M. L., Chung, S. H., Slack, A. C., et al. (1996). Isolation of human and murine homologues of the Drosophila minibrain gene: Human homologue maps to 21q22.2 in the Down syndrome "critical region" *Genomics*, *38*(3), 331–339.

Soppa, U., Schumacher, J., Florencio Ortiz, V., Pasqualon, T., Tejedor, F. J., & Becker, W. (2014). The Down syndrome-related protein kinase DYRK1A phosphorylates p27 (Kip1) and cyclin D1 and induces cell cycle exit and neuronal differentiation. *Cell Cycle*, *13*(13), 2084–2100.

Soundararajan, M., Roos, A. K., Savitsky, P., Filippakopoulos, P., Kettenbach, A. N., Olsen, J. V., et al. (2013). Structures of Down syndrome kinases, DYRKs, reveal mechanisms of kinase activation and substrate recognition. *Structure*, *21*(6), 986–996.

Spano, D., Heck, C., De Antonellis, P., Christofori, G., & Zollo, M. (2012). Molecular networks that regulate cancer metastasis. *Seminars in Cancer Biology*, *22*(3), 234–249.

Spillantini, M. G., Schmidt, M. L., Lee, V. M., Trojanowski, J. Q., Jakes, R., & Goedert, M. (1997). Alpha-synuclein in Lewy bodies. *Nature*, *388*(6645), 839–840.

Stoll, C., Alembik, Y., Dott, B., & Roth, M. P. (1998). Study of Down syndrome in 238,942 consecutive births. *Annales de Génétique*, *41*(1), 44–51.

Sugden, S. M., Bego, M. G., Pham, T. N., & Cohen, E. A. (2016). Remodeling of the host cell plasma membrane by HIV-1 Nef and Vpu: A strategy to ensure viral fitness and persistence. *Viruses*, *8*(3), 67.

Sugeno, N., Takeda, A., Hasegawa, T., Kobayashi, M., Kikuchi, A., Mori, F., et al. (2008). Serine 129 phosphorylation of alpha-synuclein induces unfolded protein response-mediated cell death. *Journal of Biological Chemistry*, *283*(34), 23179–23188.

Sugimoto, M., Kuwata, S., Kurishima, C., Kim, J. H., Iwamoto, Y., & Senzaki, H. (2015). Cardiac biomarkers in children with congenital heart disease. *World Journal of Pediatrics*, *11*(4), 309–315.

Sureau, C., & Negro, F. (2016). The hepatitis delta virus: Replication and pathogenesis. *Journal of Hepatology*, *64*(1 Suppl.), S102–S116.

Taylor, P., Fangusaro, J., Rajaram, V., Goldman, S., Helenowski, I. B., MacDonald, T., et al. (2012). REST is a novel prognostic factor and therapeutic target for medulloblastoma. *Molecular Cancer Therapeutics*, *11*(8), 1713–1723.

Tejedor, F., Zhu, X. R., Kaltenbach, E., Ackermann, A., Baumann, A., Canal, I., et al. (1995). Minibrain: A new protein kinase family involved in postembryonic neurogenesis in Drosophila. *Neuron*, *14*(2), 287–301.

Toiber, D., Azkona, G., Ben-Ari, S., Toran, N., Soreq, H., & Dierssen, M. (2010). Engineering DYRK1A overdosage yields Down syndrome-characteristic cortical splicing aberrations. *Neurobiology of Disease*, *40*(1), 348–359.

Trave, G., & Zanier, K. (2016). HPV-mediated inactivation of tumor suppressor p53. *Cell Cycle*, 1–2.

Vaseva, A. V., Marchenko, N. D., Ji, K., Tsirka, S. E., Holzmann, S., & Moll, U. M. (2012). p53 opens the mitochondrial permeability transition pore to trigger necrosis. *Cell*, *149*(7), 1536–1548.

Vazquez-Higuera, J. L., Sanchez-Juan, P., Rodriguez-Rodriguez, E., Mateo, I., Pozueta, A., Frank, A., et al. (2009). DYRK1A genetic variants are not linked to Alzheimer's disease in a Spanish case-control cohort. *BMC Medical Genetics*, *10*, 129.

Verollet, C., Le Cabec, V., & Maridonneau-Parini, I. (2015). HIV-1 infection of T lymphocytes and macrophages affects their migration via Nef. *Frontiers in Immunology*, *6*, 514.

Vihma, H., Luhakooder, M., Pruunsild, P., & Timmusk, T. (2016). Regulation of different human NFAT isoforms by neuronal activity. *Journal of Neurochemistry*, *137*(3), 394–408.

Vincent, H. A., Ziehr, B., & Moorman, N. J. (2016). Human cytomegalovirus strategies to maintain and promote mRNA translation. *Viruses*, *8*(4), e97.

Wales, T. E., Hochrein, J. M., Morgan, C. R., Emert-Sedlak, L. A., Smithgall, T. E., & Engen, J. R. (2015). Subtle dynamic changes accompany Hck activation by HIV-1 Nef and are reversed by an antiretroviral kinase inhibitor. *Biochemistry*, *54*(41), 6382–6391.

Wang, P., Alvarez-Perez, J. C., Felsenfeld, D. P., Liu, H. T., Sivendran, S., Bender, A., et al. (2015). A high-throughput chemical screen reveals that harmine-mediated inhibition of DYRK1A increases human pancreatic beta cell replication. *Nature Medicine*, *21*(4), 383.

Wang, Y., Shi, M., Chung, K. A., Zabetian, C. P., Leverenz, J. B., Berg, D., et al. (2012). Phosphorylated alpha-synuclein in Parkinson's disease. *Science Translational Medicine*, *4*(121), 121ra120.

Wegiel, J., Kaczmarski, W., Barua, M., Kuchna, I., Nowicki, K., Wang, K. C., et al. (2011). Link between DYRK1A overexpression and several-fold enhancement of neurofibrillary degeneration with 3-repeat tau protein in Down syndrome. *Journal of Neuropathology and Experimental Neurology*, *70*(1), 36–50.

Werness, B. A., Levine, A. J., & Howley, P. M. (1990). Association of human papillomavirus types 16 and 18 E6 proteins with p53. *Science*, *248*(4951), 76–79.

Wiechmann, S., Czajkowska, H., de Graaf, K., Grotzinger, J., Joost, H. G., & Becker, W. (2003). Unusual function of the activation loop in the protein kinase DYRK1A. *Biochemical and Biophysical Research Communications*, *302*(2), 403–408.

Wolfe, M. S. (2009). Tau mutations in neurodegenerative diseases. *Journal of Biological Chemistry*, *284*(10), 6021–6025.

Woods, Y. L., Rena, G., Morrice, N., Barthel, A., Becker, W., Guo, S., et al. (2001). The kinase DYRK1A phosphorylates the transcription factor FKHR at Ser329 in vitro, a novel in vivo phosphorylation site. *Biochemical Journal*, *355*(Pt. 3), 597–607.

Wu, A., Ericson, K., Chao, W., & Low, W. C. (2010). NFAT and AP1 are essential for the expression of a glioblastoma multiforme related IL-13Ra2 transcript. *Cellular Oncology*, *32*(5–6), 313–329.

Wurstle, M. L., Laussmann, M. A., & Rehm, M. (2012). The central role of initiator caspase-9 in apoptosis signal transduction and the regulation of its activation and activity on the apoptosome. *Experimental Cell Research*, *318*(11), 1213–1220.

Xanthoudakis, S., Viola, J. P., Shaw, K. T., Luo, C., Wallace, J. D., Bozza, P. T., et al. (1996). An enhanced immune response in mice lacking the transcription factor NFAT1. *Science*, *272*(5263), 892–895.

Xu, X., Yang, D., Ding, J. H., Wang, W., Chu, P. H., Dalton, N. D., et al. (2005). ASF/SF2-regulated CaMKIIdelta alternative splicing temporally reprograms excitation-contraction coupling in cardiac muscle. *Cell*, *120*(1), 59–72.

Yabut, O., Domogauer, J., & D'Arcangelo, G. (2010). Dyrk1A overexpression inhibits proliferation and induces premature neuronal differentiation of neural progenitor cells. *Journal of Neuroscience*, *30*(11), 4004–4014.

Yang, E. J., Ahn, Y. S., & Chung, K. C. (2001). Protein kinase Dyrk1 activates cAMP response element-binding protein during neuronal differentiation in hippocampal progenitor cells. *Journal of Biological Chemistry*, *276*(43), 39819–39824.

Yao, J., Qin, X., Zhu, J., & Sheng, H. (2016). Dyrk1A-ASF-CaMKIIdelta signaling is involved in valsartan inhibition of cardiac hypertrophy in renovascular hypertensive rats. *Cardiology*, *133*(3), 198–204.

Ye, Q., Jiang, J., Zhan, G., Yan, W., Huang, L., Hu, Y., et al. (2016). Small molecule activation of NOTCH signaling inhibits acute myeloid leukemia. *Scientific Reports*, *6*, 26510.

Yin, X., Jin, N., Gu, J., Shi, J., Zhou, J., Gong, C. X., et al. (2012). Dual-specificity tyrosine phosphorylation-regulated kinase 1A (Dyrk1A) modulates serine/arginine-rich protein

55 (SRp55)-promoted Tau exon 10 inclusion. *Journal of Biological Chemistry, 287*(36), 30497–30506.

Yoon, C. H., Kim, S. Y., Byeon, S. E., Jeong, Y., Lee, J., Kim, K. P., et al. (2015). p53-derived host restriction of HIV-1 replication by protein kinase R-mediated Tat phosphorylation and inactivation. *Journal of Virology, 89*(8), 4262–4280.

Zaslavsky, A., Chou, S. T., Schadler, K., Lieberman, A., Pimkin, M., Kim, Y. J., et al. (2013). The calcineurin-NFAT pathway negatively regulates megakaryopoiesis. *Blood, 121*(16), 3205–3215.

Zeidler, S., Meckbach, C., Tacke, R., Raad, F. S., Roa, A., Uchida, S., et al. (2016). Computational detection of stage-specific transcription factor clusters during heart development. *Frontiers in Genetics, 7*, 33.

Zhang, Q., Cao, L. Y., Cheng, S. J., Zhang, A. M., Jin, X. S., & Li, Y. (2015). p53-induced microRNA-1246 inhibits the cell growth of human hepatocellular carcinoma cells by targeting NFIB. *Oncology Reports, 33*(3), 1335–1341.

Zhang, D., Li, Y., Wang, R., Li, Y., Shi, P., Kan, Z., et al. (2016). Inhibition of REST suppresses proliferation and migration in glioblastoma cells. *International Journal of Molecular Sciences, 17*(5), E664.

Zhang, Y., Liao, J. M., Zeng, S. X., & Lu, H. (2011). p53 downregulates Down syndrome-associated DYRK1A through miR-1246. *EMBO Reports, 12*(8), 811–817.

Zhang, L., Xie, D., Chen, X., Hughes, M. L., Jiang, G., Lu, Z., et al. (2016). P53 mediates colistin-induced autophagy and apoptosis in PC-12 cells. *Antimicrobial Agents and Chemotherapy*.

Zhu, H., & Lo, H. W. (2010). The human glioma-associated oncogene homolog 1 (GLI1) family of transcription factors in gene regulation and diseases. *Current Genomics, 11*(4), 238–245.

Zucchini, S., Pittaluga, A., Brocca-Cofano, E., Summa, M., Fabris, M., De Michele, R., et al. (2013). Increased excitability in tat-transgenic mice: Role of tat in HIV-related neurological disorders. *Neurobiology of Disease, 55*, 110–119.

AUTHOR INDEX

Note: Page numbers followed by "*f*" indicate figures, and "*t*" indicate tables.

A

Aarsland, D., 155
Abagyan, R., 13–14
Abbruzzetti, S., 62–63, 75–76
Abel, R., 29–30, 34–35, 47
Abrahamsson, I., 84
Acheson, R., 147
Achsel, T., 49
Ackermann, A., 131
Acquafondata, M., 145–146
Adam, D., 137
Adam-Klages, S., 137
Adams, G.W., 112
Adams, M.D., 109
Affabris, E., 133
Aguayo-Ortiz, R., 5–6*t*, 18–19
Ahn, S.I., 135
Ahn, Y.S., 157
Aiello, F., 145–146
Ak, P., 140–141
Ala-aho, R., 82–83
Alam, M., 72
Alamares-Sapuay, J.G., 132–133
Alberti, A., 85
Albuquerque, D.M., 147–148
Alcami, J., 133
Aldawsari, F.S., 5–6*t*, 18–19
Alembik, Y., 136–137
Alfonsi, R., 145–146
Alford, K., 147–148
Ali, A., 122
Ali, M., 133
Allan, L.A., 139, 142
Allavena, P., 135
Allen, S.W., 107–108
Allen, T.D., 144
Allsop, D., 29
Altafaj, X., 128–129
Altankov, G., 82–101
Alvarado, J.J., 132–133
Alvarez, M., 128–129, 153
Alvarez-Perez, J.C., 157
Amadei, A., 63, 76
Amelio, I., 139–140
Anand, A., 70
Anderson, S., 62–63
Andreadis, A., 151–152
Anfinrud, P., 62–63
Annicchiarico-Petruzzelli, M., 139–140
Anselmi, M., 63, 76
Anwar, T., 139–140
Aparicio, C., 83–85, 94–95
Araghi, R.R., 28, 41
Arakawa, H., 139–140
Aranda, S., 128–129, 139, 142
Arbones, M.L., 155–156
Arguelles, A.O., 2
Arima, Y., 95–96
Arimondo, P.B., 2–3
Arisawa, A., 106, 120–121
Arkhipov, A., 63
Arkin, M.R., 28–29
Armand, A.S., 138
Arrar, M., 63–64
Arron, J.R., 134, 137, 149, 157
Arrowsmith, C.H., 2
Arroyo-Máñez, P., 62–63
Artym, V.V., 82–83
Arya, S., 70
Ascenzi, P., 62–63, 66, 72
Ashburn, T.T., 19
Atkins, K.M., 133
Audia, J.E., 2
Audie, J., 29–30
Aumailley, M., 82–83
Austin, C., 2
Axelrod, J., 106
Azkona, G., 140–141
Azorsa, D.O., 149–151
Azzarito, V., 28, 41

B

Badouel, C., 142–144
Baek, K.H., 140–141, 147

Baell, J., 2
Bagni, C., 48–49
Baier, A., 145–146
Baird, C.L., 29–30
Bajetto, A., 19
Baker, D., 36
Ballard, C., 155
Ballester-Beltrán, J., 83–84
Ballesteros-Yanez, I., 155–156
Baltimore, D., 133
Bamberg-Lemper, S., 146–147
Banko, J.L., 42–43, 48–49
Banks, J.L., 36
Banushkina, P., 63
Barallobre, M.J., 139, 142
Barbieri, F., 19
Barczyk, M., 84–85
Bardou-Jacquet, E., 132–133
Barhoum, R., 139, 142
Barron, D.A., 142–144
Barthel, A., 157
Barua, M., 151–152
Bashford, D., 46
Basmaciogullari, S., 133
Basran, A., 113
Basse, M.J., 28–29
Baudot, A.D., 139–140
Baum, B., 131–132
Baumann, A., 131
Baumgartel, K., 156–157
Baxter, L.L., 145–146
Bayo-Fina, J.M., 145–146
Beachy, P.A., 145–146
Becker, W., 128–129, 140–141, 145–149, 157
Bego, M.G., 133
Beijnen, J.H., 2–3
Bellis, L.J., 4
Bellott, M., 46
Belyi, V.A., 140–141
Ben-Ari, S., 140–141
Benavides-Piccione, R., 155–156
Benazzoug, Y., 138
Bender, A., 157
Bennett, J., 2
Bento, A.P., 4
Beran, B., 36
Berezovska, G., 76
Berg, D., 155
Berger, T., 141
Berglundh, T., 84
Bernad, S., 74
Bernard, P., 15
Bernassola, F., 140–141
Bernhardt, R., 106, 110–112, 114
Bershadsky, A.D., 83–84
Best, R.B., 46, 76
Bestor, T.H., 12
Betzi, S., 28–29
Beuerman, R.W., 42–43, 48–49
Beuming, T., 36
Bezdan, D., 139, 147
Bharate, S.B., 134
Bhutani, I., 4
Bi, C., 36
Bichet, A., 106, 110–112, 114
Bickerton, G.R.J., 28–29
Bidon-Chanal, A., 60–76
Bier, E., 129–130
Bigner, D.D., 145–146
Bigner, S.H., 145–146
Bikiel, D.E., 62–63, 66, 72–73
Binder, L.I., 149–151
Birger, Y., 139
Birsner, A., 140–141
Birukou, I., 65–66
Bishop, J.M., 144
Bisio, A., 140–141
Black, G.W., 106–123
Blagg, J., 2
Blasius, R., 135
Blaszczyk, M., 38
Bliss-Moreau, M., 147–148
Blouin, G.C., 64–65, 67
Bluhm, W.F., 36
Blumberger, J., 76
Blundell, T.L., 28–29
Bocahut, A., 74
Bodo, G., 62–63
Bodycombe, N.E., 10–12
Boechi, L., 62–64, 70–73
Boelens, R., 37, 49
Bol, S.M., 133–134
Bolognesi, M., 62–63, 66, 68–74
Bonvin, A.M., 28–30, 37, 40–41, 49
Booiman, T., 133–134

Boominathan, L., 140–141
Bordo, D., 66
Boresch, S., 30–32
Boron, I., 71–72, 75–76
Borrello, M.T., 41–42
Borst, A., 131
Bossa, C., 63
Bosveld, F., 142–144
Bouchier-Hayes, L., 139–140
Bourdon, J.C., 139–140
Bourgeas, R., 28–29
Bourne, P.E., 36
Bouzidi, S., 28–29
Bowman, J.D., 2
Bowman, L.A., 71–72
Boy, R.G., 16, 18
Boyce, S.E., 33–34
Boyl, P.P., 48–49
Bozza, P.T., 137, 140–141
Bradner, J., 154
Brandt, T., 140–141
Branson, K., 32–34
Braun, R., 46, 63
Braun, W., 107–108
Brautigam, G.R., 149–151
Bregere, C., 131
Brenke, R., 36
Brenner, D., 141
Briggs, M.M., 137–138
Britt, C., 140–141, 147
Brocca-Cofano, E., 135
Brodie, B.B., 106
Brotchi, J., 144
Brown, C.J., 42–43, 48–49
Brown, J.C., 84–85
Brown, S.P., 35–36
Brown, W.M., 151–152
Brueckner, B., 16, 18
Bruno, S., 75–76
Brunori, M., 63, 74
Bruskin, A., 139–140
Bruyere, C., 144
Bryant, S.H., 134
Bryant, S.J., 83–84
Buldyrev, S.V., 39–41
Burlingame, A.L., 112
Burmester, T., 74
Busciglio, J., 144
Bush, B.L., 30–34
Bustamante, J.P., 70–72
Buxbaum, J.D., 48–49
Byeon, S.E., 132–133
Byers, D., 131
Byrnes, M.E., 44

C

Caballero, J., 46
Cafferkey, C., 145–146
Cahalan, M., 44
Calvert, V., 132–133
Campbell, L.P., 115–117
Campos, S., 135
Canal, I., 131
Candi, E., 140–141
Canty-Laird, E.G., 101
Cao, L.Y., 139–140
Capece, L., 60–76
Capoferri, L., 113–114
Carracedo, S., 84–85
Carrinski, H.A., 10–12
Casas, C., 139
Case, D.A., 39, 42
Castillo, R., 15
Cavaignac, S.M., 112
Cavalli, A., 33
Cayzer, J., 83–84
Cazade, P.A., 76
Celia-Terrassa, T., 144
Chai, J., 141
Chaikuad, A., 132–133
Chambers, J., 4
Chandy, K.G., 44
Chang, C.A., 35–36
Chang, C.E., 32–33
Chang, C.P., 134, 137, 149, 157
Chang, H.S., 139–140, 142, 146–147
Chang, K.T., 131
Chang, S.C., 34–35, 43–44
Chao, W., 137, 140–141
Chapman, S.K., 115–117
Charnock, S.J., 106–123
Charonis, A., 85
Chau, I., 145–146
Cheatham, T. E. III, 39, 42
Cheatham, T.E., 32–33
Chelli, R., 33

Chen, C.-K., 131
Chen, H., 29–30
Chen, K., 46
Chen, L., 154–155
Chen, S.J., 5–6*t*, 16
Chen, W., 35–36, 154–155
Chen, X., 139–140
Chen, Y., 4, 44–45
Chen, Y.R., 140, 142
Cheng, G., 133
Cheng, Q., 109
Cheng, S.J., 139–140
Cheng, X., 45
Chen-Hwang, M.C., 157
Cherifi, M., 138
Chetrit, B., 28–29
Chhabra, S., 34–35, 43–44
Chiarelli, L.R., 41–42
Chien, S., 129–130
Chillemi, G., 49
Chipot, C., 32, 34–35, 47–48
Chipuk, J.E., 139–140
Chittenden, T., 139–140
Chlon, T., 147–148
Cho, H.S., 62–63
Chodera, J.D., 29–30, 32–34
Chohan, M.O., 151–152
Choi, H.K., 144
Choi, O.R., 139–140
Choi, Y., 142
Chollet, M., 62–63
Chong, L., 32–33
Chou, S.T., 147–148
Choudhury, M., 2
Christofori, G., 144–145
Chu, P.H., 136–137
Chung, K.A., 154–155
Chung, K.C., 144, 156–157
Chung, S.H., 139
Cicek, A.E., 48–49
Ciossani, G., 41–42
Ciribilli, Y., 140–141
Clark, J., 118
Clarke, P.R., 139, 142
Clayton, R.A., 109
Cleghon, V., 128–129, 131
Clementi, C., 76
Clemons, P.A., 10–12
Cludts, S., 144
Coelho, N.M., 82–101
Cohen, E.A., 133
Cohen, J., 63
Cohen, M.J., 146–147
Cohen, S.M., 10
Coiras, M., 133
Colacino, J.M., 66
Cole, J.C., 36
Collard, D.M., 83–84, 86
Colomer, J.M., 136–137
Colonna, G., 142
Comeau, S.R., 36
Commandeur, J.N., 113–114
Congia, S., 67
Conway, K.L., 134–135
Cook, D.J., 106–123
Cooper, J.R., 106
Cornall, A., 132–133
Corsi, M.M., 135
Costantini, S., 142
Costanzo, A., 139–140
Coulonges, C., 133–134
Coulson, R., 109
Coutandin, D., 140–141
Couture, M., 66
Cox, T.R., 83
Crabtree, G.R., 140–141
Crawford, M.J., 137–138
Crespo, A., 66, 68–69
Crighton, D., 139–140
Crowley, S., 147
Crowther, R.A., 151–152
Cui, M., 36
Cunha, A.F., 147–148
Czajkowska, H., 128–129, 145–146

D

D'Annessa, I., 48–49
D'Auria, K.M., 36
da Costa Martins, P.A., 138
Dabbs, E.R., 113
Dagliyan, O., 36
Daiber, A., 114
Daidone, I., 63
Daley, W.P., 82–84
Dalton, N.D., 136–137
D'Arcangelo, G., 147

Darden, T., 39, 42
Das, A.A., 37–38
Das, J., 134
Dashdorj, N., 62–63
Davidge, K.S., 71–72
Davies, M., 4
Day, Y.S., 29–30
De Antonellis, P., 144–145
de Graaf, K., 128–129, 145–146
de la Luna, S., 128–129, 153
de Lagran, M.M., 155–156
De Laurenzi, V., 139–140
De Michele, R., 135
de Ruiter, A., 32–36
de Visser, S.P., 117–118
De Voss, J.J., 112
de Vries, S. J., 39
Deadman, J.J., 49–51
DeCaprio, J.A., 147
Decker, C.J., 113–114
Degoutin, J.L., 142–144
Dehejia, A., 154
Del Prete, A., 135
Del Rio, A., 10
Del Valle, L., 133, 135
Delabar, J.M., 138
DeMaria, C.T., 42–43, 48–49
Demmer, O., 139–140
DeNeve, N., 144
Deng, L., 29–30
Deng, N., 29–30, 34, 47, 49–51
Deng, Y., 33–35, 43–44, 47
Deobagkar, D., 5–6*t*, 18
Derissen, E.J.B., 2–3
Devi, S., 138
Dewilde, S., 62–67, 72, 74
Di Daniele, N., 139–140
Di Lella, S., 62–63
Di Marino, D., 48–49
Di Nola, A., 63
Di Vona, C., 139, 147
Dicato, M., 135
Dickey, C., 149–151
Dickman, D.K., 131
Diebold, L., 147–148
Dierck, F., 137
Dierssen, M., 140–141, 155–156
Dikshit, K.L., 67–68
Dill, K.A., 29–30, 33–34
Dimitropoulos, D., 36
Dimoudis, N., 84–85, 100–101
Ding, F., 36, 38
Ding, J.H., 136–137
Ding, S., 152
DiNola, A., 76
Dintzis, H.M., 62–63
Diot, A., 139–140
Dioum, E.M., 72
Dixon, R.W., 32–34
Djoumbou, Y., 10, 19–20
Do, Q.T., 15
Dodhia, V.R., 122
Dohmae, N., 154
Dokholyan, N.V., 36, 38–41
Dolios, G., 157
Domenech, A., 139
Dominguez, C., 37, 49
Domogauer, J., 147
Dore, V., 18
Dotsch, V., 140–141
Dott, B., 136–137
Dreyfuss, G., 151–152
Droin, N.M., 139–140
Drysdale, T.A., 137–138
Dueñas, A., 4, 16, 17*f*, 18–19
Dueñas-González, A., 4–9, 5–6*t*, 19–20
Dufau, I., 2
Dufrasne, F., 144
Dumaop, W., 135
Dumont, P., 139–140
Dunbrack, R. L., Jr., 46
Duncan, G.S., 141
Dunford, A.J., 115–117
Dunkerson, J., 137
Dupont-Gillain, C., 83–84
Durand, D.B., 140–141
Duray, A., 144
Dushay, M.S., 132
Dutra, A., 154

E

Ebert, U., 84–85, 100–101
Eble, J., 84–85
Eirmbter, K., 128–129
Ekanem, I.S., 115–117
el Azzouzi, H., 138

El-Agnaf, O., 154–155
El-Deiry, W.S., 139–140
Eldon, E.D., 132
Eleuteri, B., 48–49
Eleuteri, S., 135
Emert-Sedlak, L.A., 132–133
Emmel, E.A., 140–141
Emmrich, S., 147–148
Engel, E., 83–86, 94–95
Engen, J.R., 132–133
Eramian, D., 38
Erdmann, A., 2–3
Ericson, K., 137, 140–141
Erler, J.T., 83
Errachid, A., 84–86
Erster, S., 139–140
Espinoza, I., 145–146
Estarellas, C., 60–76
Estivill, X., 128–129, 153
Estrín, D.A., 60–76
Eswar, N., 38
Ethayathulla, A.S., 140–141
Evans, N.D., 83–84
Evanseck, J.D., 46
Ewald, A.J., 83–85
Ewen, K.M., 106, 110–112, 114
Ewing, T. J. A., 36

F

Fabris, M., 135
Fahy, J., 2–3
Fan, K., 155–156
Fangusaro, J., 146
Fantuzzi, A., 122
Farahani, R., 141
Feany, M.B., 154–155
Febbraro, F., 154–155
Feig, M., 46
Feld, M.K., 83–84
Felsenfeld, D.P., 157
Felts, A.K., 36
Fenimore, P.W., 60
Fernandes, K., 139–140
Fernandez, E., 139, 142
Fernández-de Gortari, E., 2–22, 5–6*t*
Fernandez-Martinez, J., 145–146
Fernandez-Martinez, P., 139
Ferraresi, R., 144
Field, M.J., 46
Fields, J., 135
Filippakopoulos, P., 128–129
Finzel, B.C., 107–109, 108–109*f*
Firth, R., 147
Fischbach, K.F., 129–131
Fitzgerald, M., 29–30
Fleischmajer, R., 100–101
Fleischmann, R.D., 109
Flitsch, S.L., 113
Floden, E.W., 83–84
Florencio Ortiz, V., 147
Florens, L.A., 147
Florio, T., 19
Folmer, F., 135
Forli, S., 29–30, 34, 36, 47, 49–51
Forti, F., 70, 72–73
Fotaki, V., 139, 142, 155–156
Frank, A., 149–151
Frank, D., 137
Frantz, C., 82–83
Frasca, A., 42
Frauen, R., 137
Frauenfelder, H., 60
Freilino, M.L., 145–146
Freitas, T.A.K., 72
Frey, N., 136–137
Freyss, J., 10
Fricker, L.D., 29
Friedman, P., 139–140
Friesner, R.A., 36
Frost, D., 149–151
Fujii, T., 106
Fujita, K., 139–140, 142
Fujiwara, H., 154
Fujiwara, K., 82–83
Fukamizu, A., 114
Fukushima, S., 10
Fulcher, S.R., 117–118
Fulcher, T., 147
Fulco, A.J., 113–114
Fumarulo, R., 135
Funakoshi, E., 136–137
Furman, J.L., 137

G

Gaglia, M.M., 132
Gagnon, J.D., 134–135

Gai, W.P., 154–155
Galceran, J., 128–129, 145–148
Gallicchio, E., 28–51
Gallo, J.M., 135
Gan, W., 33, 47–48
Ganellin, C.R., 12
Ganguli, B.N., 107
Gao, X., 134, 137, 149, 157
Gao, Y., 29–30
García, A.J., 83–84, 86
Gardner, A.M., 67–68
Gardner, P.R., 67–68
Garg, A., 142–144
Gaudette, L., 106
Gaulton, A., 4
Gautam, V., 142
Gayraud, B., 82–83
Gazur, B., 115–117
Geerke, D.P., 113–114
Geiger, B., 83–84
Gelbcke, M., 144
Geleziunas, R., 29–30, 49–51
Gelpí, J.L., 68–70
Genheden, S., 35–36
George, D.L., 139–140
George, J.H., 83–84
Gerona-Navarro, G., 42
Gervasio, F.L., 33
Geuens, E., 66
Ghirga, F., 145–146
Gibson, Q.H., 64–65
Gilardi, G., 122
Gilson, M.K., 4, 30–36
Giovenzana, C., 144
Girhard, M., 120–121
Girvan, H.M., 113–117
Giulianotti, M.A., 10, 19–20
Given, J.A., 30–34
Gladka, M.M., 138
Gladwin, M.T., 60
Glenewinkel, F., 146–147
Godinho, F.J., 147–148
Goedert, M., 151–152, 154
Goel, G., 134–135
Goff, P.H., 132–133
Gohlke, H., 39, 42
Golbik, R., 84–85
Goldberg, A.P., 48–49
Goldberg, M.S., 154
Goldman, S., 146
Gong, C.X., 151–153
Gong, J., 139–140
González, G., 72
Gonzalez-Garcia, C., 84–87, 91–92, 94–96
Gooderham, N.J., 10
Goodsell, D.S., 36
Goody, M.F., 82–83
Gorbatyuk, O.S., 154–155
Gorr, T.A., 66
Gotoh, O., 114
Graber, T., 62–63
Graveley, B.R., 151–152
Graves, A.P., 29–30, 33–34
Gray, J.J., 36
Gray, N.S., 132–133
Green, B.N., 74
Green, J.A., 82–83
Green, M.T., 115–117
Greenway, A., 132–133
Gregoire, J.-M., 2
Greter, A., 113
Gribskov, M., 129–130
Gricman, L., 115
Griffin, M., 147
Grinnell, F., 83–84, 94–95, 100–101
Grogan, G., 107, 113
Groom, C.R., 28–29
Grootenhuis, P.D., 113–114
Gros, C., 2, 18
Groth, T., 83–85, 94–95, 100–101
Grotzinger, J., 145–146
Grundke-Iqbal, I., 149–153
Gu, D., 145
Gu, J., 153
Gu, X., 151–152
Gu, Y., 141
Guallar, V., 65–66
Guendel, I., 132–133
Guengerich, P.F., 109
Guertin, M., 66, 70
Gugutkov, D., 84–85
Guha, R., 10
Guianvarc'h, D., 2–3
Guimera, J., 139
Gullberg, D., 84–85

Gumbart, J., 46
Gumbart, J.C., 34, 47–48
Gunsalus, I., 107
Gunsalusll, I.C., 107–109, 108–109*f*
Guo, A.C., 10, 19–20
Guo, S., 157
Guo, X., 139, 142, 157
Gupta, P., 4
Gurbuxani, S., 147–148
Gustavsson, J., 83–86, 94–95
Guvench, O., 32–33
Guyader, D., 132–133
Gwack, Y., 131, 149, 157

H

Hafey, M., 139–140
Hagn, F., 139–140
Hakem, A., 141
Hakem, R., 141
Halby, L., 2–3
Halder, G., 142–144
Halgren, T.A., 36
Hall, G., 147–148
Hallermeier, K., 137–138
Halpert, J.R., 107–108
Hamdane, D., 74
Hammerle, B., 156–157
Hamon, V., 28–29
Han, B., 29–30, 49–51
Hancock, A.M., 154
Hankeln, T., 62–63, 74
Hannemann, F., 106, 110–112, 114
Hansen, H.D., 154–155
Haq, I., 112
Harder, E., 29–30
Hardy, J., 29
Harford-Cross, C.F., 117–118
Harman, J., 82–83
Hartshorn, M.J., 36
Harvey, K., 142–144
Hasegawa, M., 153–154, 157
Hasegawa, T., 154–155
Hasle, H., 139, 147
Hauer, B., 121
Haupt, S., 139–140
Haupt, Y., 139–140
Hauser, K.F., 133, 135
Hawkes, D.B., 112
Hayashi, K., 120
Haydar, T.F., 141
He, J., 136–137
He, P., 29–30, 34, 47, 49–51
He, X., 48–49
Hecht, S.S., 10
Heck, C., 144–145
Heino, J., 84–85
Heinzelmann, G., 34–35, 43–44
Heisenberg, M., 129–131
Heitmann, K., 147–148
Hekerman, P., 147–148, 156–157
Helenowski, I.B., 146
Henderson, J.T., 141
Henning, R., 62–63
Henry, C.A., 82–83
Heo, L., 37–38
Herbein, G., 132
Hersey, A., 4
Hetzer, R., 136–137
Hibshoosh, H., 139–140
Higueruelo, A.P., 28–29
Hikasa, H., 145
Hill, A., 101
Hille, S., 137
Himpel, S., 128–129
Hindupur, A., 107–108
Hoch, B., 136–137
Hochrein, J.M., 132–133
Hoffman, H., 121
Hollanda, L.M., 147–148
Holz, P., 145–146
Holzmann, S., 139–140
Hong, B., 139–140
Hong, J.Y., 145
Hong, S.W., 139–140
Honig, B., 29–30
Hoogewijs, D., 66
Hook, E.B., 148–149
Hoshimo, T., 114
Hou, L., 42–43, 48–49
Hou, S., 72
Hou, T., 46
Houghten, R.A., 10, 19–20
Howley, P.M., 140
Hu, W., 84, 140–141
Hu, Y., 145–146
Huang, H.H., 131–132

Huang, L., 145–146
Huang, Y., 157
Huang, Z., 4
Hubbell, J.A., 83
Hudson, B.G., 84–85, 100–101
Hughes, M.L., 139–140
Hung, C.H., 133
Hung, M., 29–30, 49–51
Huo, S., 32–33
Huq, R., 34–35, 43–44
Hur, D.Y., 142
Hwang, H., 36
Hyman, B.T., 154
Hynes, R.O., 82–85

I

Ichikawa, H., 62–63
Ichiyanagi, K., 62–63
Ide, S.E., 154
Ide, T., 136–137
Ikushiro, S., 120
Im, E., 156–157
Inestrosa, N.C., 145
Infante, P., 145–146
Ingelman-Sundberg, M., 110–112
Ionescu, A., 144
Iordanskiy, S., 132–133
Iozzo, R.V., 100–101
Iqbal, K., 149–153
Irvine, S.M., 83–84
Ishimori, K., 117–118
Islam, A.B., 139, 147
Ito, F., 136–137
Itoh, M., 120–121
Iuga, A., 131, 149, 157
Ivaska, J., 83–85
Iwama, H., 139–140, 142
Iwamoto, Y., 137–138
Iwata, H., 95–96
Iwatsubo, T., 153–154, 157
Izraeli, S., 139

J

Jabin, I., 144
Jackson, C.J., 113
Jackson, T.A., 62–63
Jager, A., 134–135
Jain, K., 4
Jakes, R., 151–152, 154
Jamroz, M., 38
Jan, L.Y., 131–132
Janel, N., 138
Janmey, P.A., 83
Jarosz, D., 139–140
Jaschinski, C., 137
Jaspars, M., 135
Jeevarajah, D., 49–51
Jeong, B.C., 146
Jeong, Y., 132–133
Jewison, T., 10, 19–20
Ji, C., 139
Ji, H., 145
Ji, K., 139–140
Jiande, G., 44–45
Jiang, G., 139–140
Jiang, H., 4
Jiang, J., 145–146
Jiang, L., 139–140
Jiang, W., 47
Jiang, W.M., 132–133
Jiang, X., 148–149
Jiang, Y., 44–45
Jin, J., 44–45
Jin, N., 153, 157
Jin, Q., 157
Jin, X., 153, 157
Jin, X.S., 139–140
Jo, S., 47
Joerger, A.C., 140–141
Johansson, S., 101
Johnson, M., 9–10, 83–85
Johnson, M.S., 84–85
Johnson, R.L., 142–144
Johnson, S., 132–133
Jones, A., 122
Jones, E.L., 155
Jones, G.S., 29–30, 49–51
Jones, K.D., 144
Joost, H.G., 128–129, 145–146, 157
Jorissen, R.N., 4
Joseph, J., 138
Joseph, L., 138
Joshi, M., 5–6*t*, 18
Joyce, M.G., 115–117
Juban, G., 147–148
Jun, S., 145

Jung, D.J., 139–140
Jung, H.C., 113–114

K

Kabro, A., 18
Kaczmarski, W., 151–152
Kadler, K.E., 101
Kadri, F., 133, 135
Kagawa, N., 107–108
Kagey, J.D., 142–144
Kähäri, V.-M., 82–83
Kahn, M., 145
Kalko, S.G., 68–69
Kalluri, R., 82–85, 94, 100–101
Kaltenbach, E., 131
Kaltezioti, V., 85
Kamakura, M., 120
Kamino, K., 149–151
Kan, Z., 146
Kang, J.F., 155–156
Kang, Y., 144
Kapilashrami, K., 5–6*t*, 18–19
Käpylä, J., 83–85
Karasuyama, H., 141
Karczewski, P., 136–137
Karplus, M., 30–32, 46
Kaspar, S., 146–147
Kasten-Sportes, C., 139
Katagiri, M., 107
Kataoka, N., 151–152
Kato, J., 106
Katus, H.A., 137–138
Kaus, J.W., 29–30
Kawashima, A., 154
Kay, L.J., 128–157
Kazui, H., 149–151
Keating, A.E., 28, 41
Kee, H.J., 146
Kelly, P.P., 122
Kem, W.R., 44
Kemp, B.E., 46–47
Kendrew, J.C., 62–63
Keresztes, Z., 83–84
Kerlavage, A.R., 109
Kern, A., 84–85
Kern, S.E., 139–140
Keselowsky, B.G., 83–84, 86
Kettenbach, A.N., 128–129
Keuren, M.L., 139
Khaytin, I., 44
Khor, B., 134–135
Khoshnoodi, J., 84–85, 100–101
Khosla, S., 139–140
Kida, T., 149–151
Kiger, L., 60, 66, 74
Kikuchi, A., 154–155
Kilburg, D., 28–51
Kim, B., 34–35, 47
Kim, B.-H., 36
Kim, D., 113–114, 142
Kim, D.H., 113–114
Kim, E.J., 153–154, 157
Kim, J.E., 139–140
Kim, J.H., 12–14, 137–140
Kim, K.P., 132–133
Kim, M.Y., 146
Kim, N.S., 146
Kim, S.Y., 132–133
Kim, V.N., 151–152
Kim, Y.J., 147–148
Kim, Y.S., 142
Kimura, R., 149–151
King, C.R., 146–147
Kinstrie, R., 131
Kinzler, K.W., 139–140, 145–146
Kirkness, E.F., 109
Kirsammer, G., 147–148
Kiss, R., 144
Kittaka, A., 120
Kizawa, H., 114
Klann, E., 42–43, 48–49
Klein, C., 139–140
Klicic, J.J., 36
Klimovich, P.V., 29–30
Klinedinst, D.K., 145–146
Klingenberg, M., 106
Klusmann, J.H., 147–148
Kmiecik, S., 38
Knapp, S., 132–133
Knight, Z.A., 133
Knox, C., 10, 19–20
Ko, J., 37–38
Kobayashi, K., 139–140, 142
Kobayashi, M., 154–155
Koch, M.L., 147–148
Kogerman, P., 157

Kohler, V., 122
Kolahi, K.S., 82–83
Kolinski, A., 38–39
Kollman, P.A., 32–33
Kon, N., 139–140
Kondrikova, G., 154–155
Kook, H., 146
Kootstra, N.A., 133–134
Kosik, K.S., 151–152
Kounatidis, I., 132
Kozakov, D., 36
Kragh, C.L., 154–155
Kraner, S.D., 137
Krasselt, A., 62–63
Krause, E.G., 136–137
Kraut, J., 107–109, 108–109*f*
Krieg, P.A., 137–138
Krilov, G., 34–35, 47
Krishna, R., 37–38
Kuan, C.Y., 141
Kubler, W., 137–138
Kuchna, I., 151–152
Kuchroo, V.K., 134–135
Kuck, D., 15–16, 18
Kuenemann, M.A., 28–29
Kuhl, C., 137
Kuhlman, B., 36
Kuhn, B., 32–33
Kuhn, C., 137
Kühn, K., 84–85, 100–101
Kuida, K., 141
Kumar Dhanyamraju, P., 145–146
Kumar, A., 132
Kumar, D., 63
Kumar, M.S., 37–38
Kumar, N., 132–133
Kumar, S., 107–108
Kuntz, I.D., 36, 62–63
Kurcinski, M., 38–39
Kurishima, C., 137–138
Kuwana, T., 139–140
Kuwata, S., 137–138
Kuwazaki, S., 114
Kuyucak, S., 34–35, 43–44

L

La Du, B.N., 106
Labbé, C.M., 28–29
Laberge, M., 60–61
Lachance, H., 18
Laconde, G., 28–29
Laguna, A., 139, 142
Lai, C.R., 140, 142
Lakshminarayanan, R., 42–43, 48–49
Lama, A., 67–68, 70
Lama, D., 42–43, 48–49
Lamb, D.C., 113
Lamoral-Theys, D., 144
Lane, D.P., 42–43, 48–49
Lapelosa, M., 32, 34, 48–51
Larsen, M., 82–84
Laussmann, M.A., 141
Lavedan, C., 154
Law, M.L., 145–146
Law, V., 10, 19–20
Lawson, R.J., 112
Lazaridis, T., 29
Le Cabec, V., 133
Le Grice, S.F., 45
Leder, S., 128–129
Lee, H., 37–38
Lee, H.J., 153–154, 157
Lee, H.K., 142
Lee, J., 132–133
Lee, J.H., 146
Lee, M., 145
Lee, M.S., 37–38
Lee, V.M., 154
Lefranc, F., 144
Lei, J., 44–45
Lei, L., 107–108
Leiden, J.M., 137–138
Leitgeb, M., 30–32
Lendlein, A., 83–85, 94–95
Leptidis, S., 138
Leroy, E., 154
Leu, J.I.J., 139–140
Leverenz, J.B., 155
Leveritt, J.M., 29
Levine, A.J., 140
Levrero, M., 139–140
Levy, R.M., 29–34, 36, 47–51
Li, H., 145–146
Li, K., 155–156
Li, L., 145
Li, S., 154–155

Li, S.S., 5–6*t*, 16
Li, T., 131–132, 139–140
Li, X., 139, 142, 157
Li, Y., 113–114, 139–140, 146
Liang, H., 153
Liang, J., 146
Liang, Y.J., 139, 146–147
Liang, Y.Y., 132–133
Liao, J.M., 139–141, 147
Lieberman, A., 147–148
Lighthouse, J.K., 137
Ligoxygakis, P., 132
Lim, I.K., 139–140
Lim, M., 62–63
Lima, C.S., 147–148
Limou, S., 133–134
Lin, C.H., 140, 142
Lin, T., 29–30
Lin, T.A., 134
Lin, Y.-L., 29–30
Lin, Y.M., 4
Lindberg, P., 12
Linder, E., 84
Lindhe, J., 84
Litovchick, L., 147
Liu, C., 134, 148–149
Liu, F., 151–152
Liu, G., 139–140
Liu, H., 139
Liu, H.T., 157
Liu, L., 132–133
Liu, N., 139
Liu, Q., 139
Liu, T., 131–132
Liu, T.Q., 4
Liu, X., 4, 29–30, 49–51
Liu, Y., 10, 19–20
Liu, Z.H., 155–156
Llopis-Hernández, V., 82–101
Lo Celso, C., 145–146
Lo, H.W., 145–146
Locascio, J.J., 154
Lochhead, P.A., 128–129, 131
Loharch, S., 4
Lombardo, D., 144
Lomonosova, E., 45
London, N., 28–29, 37, 40–41
Londos, E., 155
Lonez, C., 144
Long, K., 28, 41
Looser, S., 137–138
Lopes, P. E. M., 46
Lopez-Bigas, N., 139, 147
Lopez-Huertas, M.R., 133
López-Vallejo, F., 15
Lorho, R., 132–133
Loukachov, V.V., 133–134
Low, W.C., 137, 140–141
Lowe, D.J., 113
Lowe, S.W., 144
Lu, B., 46
Lu, G., 45
Lu, H., 139–141, 147
Lu, H.Z., 132–133
Lu, J.F., 131
Lu, J.Y., 136–137
Lu, P., 83
Lu, S., 137–138
Lu, Z., 139–140
Lucas, F., 65–66
Luhakooder, M., 137
Lun, S., 83–84
Lund, L.B., 154–155
Luo, C., 137, 140–141
Luo, M., 5–6*t*, 18–19
Luo, R., 39, 42
Luque, F.J., 60–76
Ly, H., 132–133
Lyko, F., 15–16, 18
Lynch, R.C., 140–141, 147

M

Macdonald, D.E., 100–101
MacDonald, T., 146
Machida, K., 120–121
MacKerell, A. D., Jr., 46
MacKerell, A.D., 32–33
Maenz, B., 147–148
Maggiora, G.M., 14, 19–20
Mahadevan, M.S., 141
Mahnir, V.M., 44
Mainz, D.T., 36
Mak, J., 132–133
Mak, T.W., 141
Makamura, N., 119
Makino, S., 36

Malca, S.H., 121
Maldonado-Rojas, W., 5–6*t*, 15
Malinge, S., 147–148
Maneva-Radicheva, L., 84–85, 100–101
Máñez, P.A., 70–71
Manfredsson, F.P., 154–155
Mangino, G., 133
Manning, N.J., 113
Mantovani, A., 135
Mao, J., 157
Mao, L., 136–137
Marchant, M.J., 139–140
Marchenko, N.D., 139–140
Marcoux-Archambault, I., 18
Marczylo, T.H., 113
Marden, M.C., 60, 66
Marechal, J.-D., 65–66
Maridonneau-Parini, I., 133
Marino, D.D., 49
Markert, E., 140–141
Maroz, A., 147–148
Marrero-Ponce, Y., 5–6*t*, 15
Martí, A.M., 70–71
Martí, M.A., 63–64, 66–75
Martin de Lassalle, E., 144
Martinez-Gil, L., 132–133
Martínez-Mayorga, K., 10, 13–14, 20–21
Martyn, J., 49–51
Maruo, F., 114
Masliah, E., 154
Massova, I., 32–33
Mateo, I., 149–151
Mathieu, V., 144
Mathur, P.P., 37–38
Matsuda, K., 139–140
Matsumura, H., 119
Matsuoka, T., 110–112
Mattevi, A., 41–42
Mattson, M.P., 141
May, A., 39
Maye, P., 157
Mazzanti, M., 19
McCammon, J.A., 29–34
McCulloch, C.A., 83
McIlwain, D.R., 141
McIntyre, K.W., 134
McKenney, R.J., 131–132
McKinsey, T.A., 136–137
McLaughlin, W.A., 46
McLean, K.J., 113–117
Mclean, M., 141
McLean, P.J., 154
McMahon, B.H., 60
McNeill, H., 142–144
McReynolds, A.C., 29–30
Means, A.R., 136–137
Meckbach, C., 137
Medina-Franco, J.L., 2–22, 5–6*t*, 17*f*
Meec hoovet, B., 149–151
Meeusen, S., 157
Meharenna, Y.T., 112
Melchert, R.B., 138
Melino, G., 139–141
Melquiond, A.S., 40–41
Meltzer, S., 131–132
Mena-Ulecia, K., 46
Méndez-Lucio, O., 2–22, 17*f*
Meng, K., 148–149
Meng, X.-Y., 36
Mercaldo, V., 48–49
Meruvu, S., 2
Merz, K., 39, 42
Meurice, N., 19–20
Meuwly, M., 63, 76
Meyer, R., 136–137
Meyers, M.J., 45
Mezei, M., 36
Miele, L., 145–146
Mihara, M., 139–140
Mijatovic, T., 144
Milani, M., 62–63, 66, 68–71
Miles, C.S., 118
Miles, J.A., 42
Miller, R.K., 145
Milton, C.C., 142–144
Mishra, S., 76
Mitoma, C., 106
Mitscher, L.A., 12
Mittal, J., 46
Mobley, D.L., 29–30, 32–34
Moens, L., 60, 62–67
Moerland, P.D., 133–134
Mofrad, M.R., 82–83
Moll, U.M., 139–140
Mollenhauer, B., 154
Montes, C.P., 20–21

Monti, P., 140–141
Moon, I.S., 84
Moon, J.H., 139–140
Moorman, N.J., 132
Moran, E.A., 45
Moratal, D., 83–84
Morceau, F., 135
Morgan, C.R., 132–133
Morgan, T.H., 129–130
Mori, F., 154–155
Mori, M., 145–146
Morishima, I.J., 117–118
Morita, K., 136–137, 145–146
Morreale, A., 68–69
Morrice, N., 128–129, 131, 157
Morris, N.P., 133
Mosher, D.F., 101
Mouawad, L., 65–66
Moughon, S., 36
Movshovitz-Attias, D., 28–29, 40–41
Mowat, C.G., 115–118
Muchmore, S.W., 35–36
Mujtaba, S., 42
Mukai, M., 70
Mukund, S., 29–30, 49–51
Muller, M., 19–20
Muller, M.R., 147–148
Muller, S., 132–133
Munoz, W.A., 145
Munro, A.W., 112–117
Murakami, N., 157
Murali, T.M., 63
Muralidhara, B.K., 107–108
Murayama, N., 106
Murdoch, A.D., 100–101
Murphy, M.E., 139–140
Murphy, N.S., 28, 41
Murphy, R.B., 36
Murray, C.W., 36
Murray-Zmijewski, F., 139–140
Musch, T., 16, 18
Mymryk, J.S., 146–147
Myszka, D.G., 29–30

N

Nadra, A.D., 74–75
Nakamura, K., 139–140
Nakamura, Y., 139–140
Nakano, K., 139–140
Napoli, I., 48–49
Nara, F., 110–112, 119
Nardini, M., 62–63, 66–67, 72–73
Nardone, J., 131, 149, 157
Narhi, L.O., 113–114
Narikawa, T., 106
Narwade, S.C., 5–6*t*, 18
Nasi, M., 144
Navani, N.K., 67–68
Navarrete-Vázquez, G., 14
Naveja, J.J., 5–6*t*, 14–15, 19–20
Neeli, R., 115–117
Negro, A., 154
Negro, F., 132
Negron, L., 83–84
Nelson, D.R., 114
Nerattini, F., 33
Nestl, B.M., 121
Neves, M.A.C., 13–14
Newhouse, J., 72
Newman, J., 49–51
Newmeyer, D., 139–140
Nguyen, H.T., 140–141
Nguyen-Tran, V., 157
Nichols, C.D., 132
Nickerson, D.P., 117–118
Nicklen, S., 109
Nicklin, S., 113
Nicolai, S., 139–140
Nienhaus, G.U., 66–67
Nienhaus, K., 62–63, 66–67
Nieto, M.A., 145–146
Niwa, T., 106
Noll, C., 138
Norton, A., 147–148
Nowicki, K., 151–152
Nozawa, S., 62–63
Nuripa, A., 149–151
Nutt, D.R., 63
Nykvist, P., 83–85

O

O'Brien, S.J., 145–146
Ocana, O.H., 145–146
Oda, M., 114
O'Day, K.D., 134
O'Donnell, J., 132–133

Oezguen, N., 107–108
Ogita, K., 136–137
Ogliaro, F., 117–118
Ohtaki, A., 119
Okada, Y., 140–141, 147
Okamure, W.H., 120
Okui, M., 136–137
Okura, R., 139–140, 142
Olivares, W., 139–140
Oliveira, A., 70
Olivero-Verbel, J., 5–6*t*, 15
Olsen, J.V., 128–129
Olson, A.J., 36
Olson, E.N., 136–137
Olson, J.S., 62–66
O'Meara, N., 147
Omi, S., 119
Onufriev, A.V., 63
Oostenbrink, C., 32–36
Oostra, B.A., 48–49
O'Prey, J., 139–140
O'Reilly, E., 122
Ori-McKenney, K.M., 131–132
Orozco, M., 68–69
Ortiz de Montellano, P.R., 112–113
Ossowski, S., 139, 147
Ost, T.W.B., 118
Ouellet, Y., 66, 70
Ozelo, M.C., 147–148

P

Pacifici, M., 133, 135
Packman, L.C., 128–129
Padovan, J.C., 157
Page-McCaw, A., 83–85
Pahl, R., 62–63
Paladino, A., 142
Palese, P., 132–133
Paluch, J., 131
Pan, J.G., 113–114
Pan, Y., 134
Panagiotis, L.K., 28–30
Panchenko, A.R., 29–30
Pancoska, P., 139–140
Pande, V.S., 32–34
Pandey, U.B., 132
Pankov, R., 83–84
Panzer, P., 128–129
Papadopoulos, C., 156–157
Parak, F., 62–63
Park, G.B., 142
Park, H., 37–38
Park, J., 144
Park, J.I., 145
Park, S.A., 135
Park, S.H., 113–114
Parker-Struckhoff, A., 133, 135
Parkesh, R., 4
Parmacek, M.S., 137–138
Parrish, R.G., 62–63
Parslow, T.G., 132–133
Pask, C.M., 42
Pasqualon, T., 147
Patel, D.J., 12
Patel, J., 42
Pathania, R., 67–68
Patterson, D., 147
Patterson, K.D., 137–138
Paul, D., 84–85, 101
Pawaria, S., 67–68, 70
Peat, T.S., 49–51
Pedchenko, V., 84–85, 100–101
Pegueroles, M., 83–85, 94–95
Peña-Castillo, A., 2–22
Peng, J.L., 5–6*t*, 16
Penhallow, B., 134
Pennacchietti, F., 75–76
Pennington, M.W., 44
Pentikäinen, O., 83–85
Peppard, T.L., 10
Perahia, D., 65–66
Percario, Z.A., 133
Perez-Olmeda, M., 133
Perez-Villanueva, J., 15
Periquet, M., 154
Perissinotti, L.L., 62–63
Perkins, J., 112
Perlish, J.S., 100–101
Perrier, V., 18
Perryman, A.L., 29–30, 34, 36, 47, 49–51
Perwitasari, O., 132–133
Pesce, A., 62–63, 66–67, 70, 74
Peskind, E.R., 154
Peters, S.B., 82–84
Petrenko, O., 139–140
Petrey, D., 29–30

Petricoin, E., 132–133
Petrovich, M., 140–141
Petsko, G.A., 62–63
Peyser, N.D., 145–146
Pham, T.N., 133
Phillips, D.C., 62–63
Phillips, G.N., 64
Phillips, J.C., 46
Piehler, J., 29–30
Pierce, B.G., 36
Pierce, L., 34–35, 47
Pilotto, S., 41–42
Pimkin, M., 147–148
Pinilla, C., 10, 19–20
Pino-Angeles, A., 29
Pirker, C., 144
Pitt, W.R., 28–29
Pittaluga, A., 135
Pivarcsi, A., 145–146
Pizzato, M., 133
Place, E.S., 83–84
Planas, A.M., 139
Planell, J., 84–86
Planell, J.A., 84–87, 91–92, 94–96
Pleiss, J., 115
Pleiss, M.M., 137
Plotnikov, A.N., 42
Poblete, H., 46
Podust, L.M., 106
Pohorille, A., 32, 34–35
Polleri, A., 134, 137, 149, 157
Polymeropoulos, M.H., 154
Potocki, L., 129–130
Potter, M.J., 35–36
Poulin, F., 42–43, 48–49
Poulos, T.L., 107–109, 108–109*f*, 112
Poultney, C.S., 48–49
Pozueta, A., 149–151
Prabhu, V.V., 139–140
Prada, N., 144
Prada-Gracia, D., 76
Premkumar Reddy, E., 132–133
Prieto-Martínez, F.D., 2–22
Prisant, M.G., 63
Prives, C., 139–140
Procacci, P., 33
Proctor, E.A., 36
Pruunsild, P., 137
Pucharcos, C., 139
Puchkaev, A.V., 112–113
Puranen, S., 84–85
Putman, E.J., 137
Puzio-Kuter, A., 140–141

Q

Qian, W., 152–153, 157
Qiang, L., 29–30
Qin, X.T., 136–137
Quah, S.T., 42–43, 48–49
Quinlan, M., 149–151
Quinn, J.F., 154

R

Raad, F.S., 137
Raaf, L., 138
Rahman, M.F., 145–146
Rajaram, V., 146
Rajpathak, S.N., 5–6*t*, 18
Ramakrishna, G., 139–140
Ramana Reddy, M.V., 132–133
Ramos, A., 140–141
Rao, A., 147–148
Rao, F., 76
Raschella, G., 139–140
Rashid, M.H., 34–35, 43–44
Ratner, B.D., 83–84
Raucy, J.L., 107–108
Rauer, H., 44
Raveh, B., 28, 37
Rechkoblit, O., 12
Reeves, R.H., 145–146
Rehm, M., 141
Reid, G.A., 118
Reinen, J., 113–114
Reinhardt, D., 147–148
Reinhardt, K., 147–148
Reinhardt, S., 74
Reiter, L.T., 129–130
Remacle, C., 83–84
Remppis, A., 137–138
Rena, G., 157
Repasky, M., 36
Reyes, C., 32–33
Reymond, J.-L., 20–21
Rhim, H., 153–154, 157
Rhodes, D.I., 49–51

Rich, R.L., 29–30
Richmond, H., 147–148
Rico, P., 83–84
Riggs, A.F., 66
Riggs, C.K., 66
Riikonen, R., 83–85
Rioth, M.J., 140–141
Risteli, J., 101
Ristorcelli, E., 144
Rittle, J., 115–117
Rivas, H.G., 132
Rizzi, M., 66
Roa, A., 137
Roat, E., 144
Roberts, G.A., 113
Roberts, I., 147–148
Roberts, S., 107
Robertson, K.D., 2, 12–14
Robeson, R.H., 149–151
Robin, A., 122
Roche, M.I., 134–135
Roche, P., 28–29
Rocklin, G.J., 33–34
Rockman, H.A., 136–137
Rodriguez, A., 139–140
Rodriguez-Marin, S., 42
Rodriguez-Rodriguez, E., 149–151
Rohl, C.A., 36
Roitberg, A.E., 63–64
Roos, A.K., 128–129
Roper, R.J., 145–146
Rose, P.W., 36
Rosser, S.J., 113
Rossi, A., 139–140
Roth, M.P., 136–137
Rousseau, D.L., 66
Roux, B., 29–30, 33–34, 43–44, 47–48
Rouxhet, P.G., 83–84
Rowell, P., 42
Roy, N., 141
Rubeis, S.D., 48–49
Ruby, C.E., 133
Ruddigkeit, L., 20–21
Rufener, C., 10
Rupasinghe, S., 107–108
Ruscio, J.Z., 63
Rutherford, D., 151–152
Ruysschaert, J.M., 144
Ryan, K.M., 139–140
Ryde, U., 35–36
Ryeom, S., 140–141
Ryu, M.S., 139–140

S

Saavedra, K., 139–140
Sabatini, S., 139–140
Sacquin-Mora, S., 74
Sahoo, D.K., 4
Saito, J.A., 72
Sakamoto, T., 139–140, 142
Saksela, K., 133
Saladino, G., 33
Sali, A., 38
Salic, K., 138
Salichs, E., 139, 147
Salmerón-Sánchez, M., 82–101
Salter, M.D., 64–65, 67
Samitier, J., 84–86
Samocha, K., 48–49
Sampey, G.C., 132–133
Sanchez-Gomez, P., 139
Sanchez-Juan, P., 149–151
Sanctis, D., 62–63
Sander, T., 10
Sanger, F., 109
Santiago, D.N., 29–30, 34, 36, 47, 49–51
Santoro, G., 135
Santos-Martins, D., 36
Saran, N.G., 145–146
Sato, T., 62–63
Saussez, S., 144
Savitsky, P., 128–129
Sayed, M., 128–129
Schachat, F., 137–138
Schadler, K., 147–148
Schechter, A., 100–101
Scheffold, T., 137–138
Schellens, J.H.M., 2–3
Scheps, A., 121
Schindler, C. E. M., 39
Schittny, J., 84–85
Schlossmacher, M.G., 154
Schmaling, S.K., 132
Schmid, R.D., 120–121
Schmidt, M., 62–63
Schmidt, M.L., 154

Schneider, P., 145–146
Schotte, F., 62–63
Schreyer, A., 28–29
Schueler-Furman, O., 28–29, 36–37, 40–41
Schug, T.T., 139, 142, 157
Schuler, M., 139–140
Schuler, M.A., 107–108
Schulten, K., 46, 63
Schulz-Schaeffer, W., 154
Schumacher, J., 147
Schwab, M., 110–112
Schweers, R.L., 65–66
Scior, T., 15
Scott, E.E., 64–65
Sebban, P., 74
Seifert, A., 139, 142
Seira, C., 60–76
Senzaki, H., 137–138
Seok, C., 37–38
Serger, E., 135
Serizawa, N., 110–112, 119
Seth-Smith, H.M.B., 113
Shaik, S., 117–118
Shakhnovich, E.I., 39–41
Shamji, A.F., 10–12
Shao, Z., 147–148
Sharma, A., 142
Sharma, N.R., 132–133
Sharma, O.P., 37–38
Sharma, S., 131, 149, 157
Shastri, V.P., 83–85, 94–95
Shaw, J.P., 140–141
Shaw, K.T., 137, 140–141
Shekhawat, N.S., 138
Shen, M.-Y., 38
Shen, W., 157
Sheng, H., 136–137
Sheng, H.Z., 136–137
Shenkin, P.S., 36
Sherman, D.H., 106
Sherman, W., 36
Shi, F., 82–83
Shi, J., 151–153, 157
Shi, M., 154–155
Shi, P., 146
Shi, Y., 29–30, 141
Shi, Z., 5–6*t*, 16
Shim, J., 46
Shimokawa, T., 145–146
Shin, J.S., 139–140
Shinkyo, R., 120
Shiozaki, E.N., 141
Shirts, M.R., 29–30, 32–34
Shoemaker, B.A., 29–30
Shoichet, B.K., 29–30, 33–34
Shoun, H., 114
Shukla, M., 63
Shutler, G., 141
Siarey, R.J., 140–141, 147
Sibbet, G., 128–129, 131
Sideraki, V., 85
Siedlecki, P., 16, 18
Sierra Gonzalez, P., 139–140
Sims, D., 131–132
Singh, N., 10, 16, 18
Singh, R., 145–146
Singh, S., 70–72
Singleton, E.W., 64–65, 67
Sippl, W., 140–141
Sitz, J.H., 156–157
Sivendran, S., 157
Sixel-Doring, F., 154
Skillman, A.G., 36
Slack, A.C., 139
Slessor, K.E., 112
Small, E.M., 137
Smirnov, A.V., 62–63
Smith, R.L., 10
Smithgall, T.E., 132–133
Smulders-Srinivasan, T.K., 128–157
Smulevich, G., 62–63
Soengas, M.S., 141, 144
Solans, A., 139
Soman, J., 62–67
Somers, J., 139–140
Sompol, P., 137
Sonenberg, N., 42–43, 48–49
Song, J., 12
Song, W.J., 139
Sonkoly, E., 145–146
Soppa, U., 147–149
Soreq, H., 140–141
Sosna, J., 137
Sottile, J., 82–83
Soundararajan, M., 128–157
Spano, D., 144–145

Spatz, J.P., 83
Spencer-Dene, B., 145–146
Sperandio, O., 28–29
Spillantini, M.G., 151–152, 154
Spitaleri, A., 33
Spyrakis, F., 62–63
Srikanth, S., 131, 149, 157
Stachorski, L., 147–148
Stahle, M., 145–146
Stamenkovic, I., 84
Stanley, H.E., 39–41
Sternberg, L.R., 139
Sternlicht, M.D., 83
Stevens, M.M., 83–84
Stewart, K.M., 82–83
Stites, W.E., 28
Stoll, C., 136–137
Su, T., 139–140
Sugden, S.M., 133
Sugeno, N., 154–155
Sugimoto, H., 120
Sugimoto, M., 137–138
Suhai, S., 16, 18
Sullivan, L.F., 154–155
Sumisa, F., 106
Summa, M., 135
Sun, L., 113–114
Sung, J.Y., 153–154, 157
Sureau, C., 132
Suzek, T.O., 4–9
Suzuki, S., 114
Swanson, J., 29–30
Swanson, S.K., 147

T

Tabudravu, J., 135
Tachedjian, G., 132–133
Tacke, R., 137
Tajhya, R.B., 34–35, 43–44
Tajkhorshid, E., 46
Takahashi, S., 117–118
Takai, K., 83
Takano, J., 139–140, 142
Takaya, N., 114
Takeda, A., 154–155
Takeda, K., 106
Tanasa, B., 131, 149, 157
Tancredi, H., 48–49
Tang, B.S., 155–156
Tang, L., 84
Tang, Y., 28–29
Tapon, N., 142–144
Tarafdar, S., 132–133
Tardugno, M., 41–42
Tarricone, C., 66
Tavis, J.E., 45
Taya, C., 141
Taya, Y., 139–140
Taylor, B., 157
Taylor, P., 146
Taylor, R.D., 36
Tejedor, F.J., 128–129, 131, 147–149, 156–157
Tejero, J., 60
Tellinghuisen, J., 29–30
Tettinger, F., 30–32
Thellung, S., 19
Thevenot, P., 84
Thijs, L., 72
Thomas, L., 133
Thor, K.B., 19
Thu, C.A., 29–30
Tilton, R.F., 62–63
Timmusk, T., 137
Timpl, R., 84–85
Tinajero-Trejo, M., 71–72
Tipping, M., 142–144
Tiznado, W., 46
Todd, E.M., 83–84
Toftgard, R., 157
Toiber, D., 140–141
Tomita, A., 62–63
Tomure, D., 114
Tong, P., 146
Tonissen, K.F., 137–138
Toole, J.J., 140–141
Toran, N., 140–141
Tora-Ponsioen, M., 145–146
Toromanov, G., 84–85
Torres-Estay, V., 139–140
Tortorici, M., 41–42
Toscano, S., 145–146
Tostar, U., 145–146
Totrov, M., 13–14
Tramontano, A., 49
Tran, J., 19–20

Tran, K., 134–135
Trave, G., 140
Travis, E.R., 113
Trejo, M., 135
Trellet, M., 40–41
Trenkwalder, C., 154
Trent, J.M., 145–146
Tripp, R.A., 132–133
Troiano, L., 144
Trojanowski, J.Q., 154
Tsao, M.S., 144
Tsiang, M., 29–30, 49–51
Tsirka, S.E., 139–140
Tubert-Brohman, I., 36
Tung, Y.C., 149–151
Turner, N.J., 113
Tzoneva, R., 84–85, 101

U

Uchida, S., 137
Ullrich, V., 114
Urlacher, V.B., 106, 120–121
Utz, P.J., 140–141
Uzan, J., 74

V

Vajda, S., 36
Valbuena, J., 139–140
Vallone, B., 63, 74
van den Heuvel, A.P., 139–140
van Dort, K.A., 133–134
Van Duyne, R., 132–133
van Leeuwen, J.S., 113–114
van Manen, D., 133–134
van Remmerden, Y., 133–134
Vandegraaff, N., 49–51
Vandergon, T.L., 66
van't Wout, A.B., 133–134
Varela-Nallar, L., 145
Vaseva, A.V., 139–140
Vassallo, J., 147–148
Vazquez-Higuera, J.L., 149–151
Vela, E.M., 145–148
Velling, T., 101
Venkataraman, H., 113–114
Veprintsev, D.B., 140–141
Verdonk, M.L., 36
Vergara-Jaque, A., 46
Verkade-Vreeker, M.C., 113–114
Verma, C.S., 42–43, 48–49
Vermeulen, N.P., 113–114
Vernier, M., 144
Verollet, C., 133
Viappiani, C., 62–63
Vihma, H., 137
Villa, E., 46
Villoutreix, B.O., 28–29
Vincent, H.A., 132
Vinogradov, S.N., 60
Viola, J.P., 137, 140–141
Vishwakarma, R.A., 134
Vlahakos, D., 85
Vogel, C., 115
von Korff, M., 10
von Wachenfeldt, C., 112
Vousden, K.H., 139–140
Vreven, T., 36

W

Wabik, J., 39
Waddell, J., 14
Waddell, W.J., 10
Wagner, B.K., 10–12
Wagner, G., 107–109, 108–109*f*
Wagner, S., 131
Wajcman, H., 60
Wakatabi, M., 119
Wales, T.E., 132–133
Walkinshaw, M.D., 118
Wallace, J.D., 137, 140–141
Waltham, T.N., 115–117
Wan, P., 107
Wang, C., 36
Wang, C.Y., 139, 146–147
Wang, H., 140–141
Wang, J., 4–9, 138
Wang, J.Y.J., 139–140
Wang, K.C., 151–152
Wang, L., 29–30, 34–35, 47, 145–146
Wang, P., 139, 157
Wang, P.-H., 76
Wang, Q., 4
Wang, R., 146, 157
Wang, S., 153, 157
Wang, S.J., 139–140
Wang, W., 46, 136–137

Wang, X., 154
Wang, Y., 4–9, 134, 155
Wang, Y.L., 5–6*t*, 16
Wang, Y.Q., 155–156
Warrilow, A.G., 113
Warriner, S.L., 42
Washburn, M.P., 147
Watanabe, I., 110–112, 119
Watt, F.M., 145–146
Weaver, V.M., 82–83
Webb, B., 38
Webb, S.P., 35–36
Weber, Y., 128–129
Wegiel, J., 151–152
Weich, B., 74
Wells, J.A., 28–29
Welmaker, G.S., 19
Wen, X., 4
Weng, Z., 36
Wennerberg, A., 84
Wennerberg, K., 101
Werb, Z., 83–85
Wermuth, C.G., 12
Werness, B.A., 140
Wetzel, K., 128
Wetzler, D.E., 75–76
White, D.J., 84–85
White, O., 109
Wiechmann, S., 145–146
Wiehe, K., 36
Wilk, A., 133, 135
Will, R., 137
Willetts, A., 107
Williams, J.G., 139, 142, 157
Wilson, A.J., 28, 41–42
Wilson, J.A., 10–12
Winslow, M.M., 134, 137, 149, 157
Wisniewski, H.M., 149–151
Wolfe, M.S., 151–152
Wong, J., 4
Wong, L.-L., 117–118
Woo, H.-J., 33, 47
Woo, M., 141
Woods, R.J., 39, 42
Woods, Y.L., 157
Wu, A., 137, 140–141
Wu, H., 134, 137, 149, 157
Wu, K., 85
Wu, Y., 34–35, 47
Wurstle, M.L., 141
Würth, R., 19
Wyckoff, H., 62–63
Wynn, T.A., 82–83

X

Xanthoudakis, S., 137, 140–141
Xia, Y., 146
Xiao, J., 4–9
Xie, D., 139–140
Xie, W., 157
Xing, Z., 148–149
Xirodimas, D.P., 139–140
Xu, J., 147–148
Xu, X., 136–137
Xu, Y.J., 9–10

Y

Yabut, O., 147
Yadav, R.R., 134
Yamada, K.M., 82–84
Yamada, M., 120
Yamaguchi, M., 114
Yamamoto, M., 149–151
Yamazaki, H., 106
Yan, W., 145–146
Yan, X.Z., 132–133
Yanagawa, N., 144
Yang, C.H., 140, 142
Yang, D., 136–137
Yang, E.J., 144, 157
Yang, L., 142–144
Yankner, B.A., 144
Yao, J., 136–137
Yaraghi, Z., 141
Yasuda, K., 120
Ye, Q., 145–146
Yeh, J.I., 132–133
Yeh, S.-R., 66, 70
Yen, M.S., 140, 142
Yeo, D.J., 42
Yin, S., 38
Yin, X., 137–138, 153, 157
Yohda, M., 119
Yoneda, S., 120
Yonetani, T., 60–61
Yoo, J., 4–9, 5–6*t*, 12–14, 16, 17*f*, 18–19

Yoon, C.H., 132–133
Yoon, J.H., 144
Yoshida, K., 139–140
Yoshioka, S., 117–118
Younker, J.M., 115–117
Yu, E., 142–144
Yu, T., 148–149
Yu, W.C., 139, 146–147
Yu, X.H., 136–137
Yuan, H., 107–108
Yurchenco, P.D., 84–85, 100–101

Z

Zabetian, C.P., 155
Zacharias, M., 39
Zahonero, C., 139
Zaika, A., 139–140
Zalfa, F., 48–49
Zang, S., 139
Zanger, U.M., 110–112
Zanier, K., 140
Zaslavsky, A., 140–141, 147–148
Zaydenberg, I., 44
Zeidler, S., 137
Zeisberg, M., 85
Zeng, L., 42
Zeng, S.X., 140
Zeng, Y., 145–146
Zhan, G., 145–146
Zhang, A.M., 139–140
Zhang, D., 146
Zhang, H.-X., 36
Zhang, J., 4–9
Zhang, L., 139–140, 146, 148–149
Zhang, Q., 139–140
Zhang, Q.C., 29–30
Zhang, S., 139–140
Zhang, T., 151–152
Zhang, X.Y., 132–133
Zhang, Y., 139–141, 147
Zhang, Y.Q., 157
Zhang, Y.Z., 132–133
Zhao, B., 107–108
Zhao, W., 146
Zheng, W., 76
Zheng, Y.G., 4–9
Zhou, C., 151–152
Zhou, J., 153
Zhou, M.-M., 42
Zhou, W., 5–6*t*, 16
Zhou, X., 139–141, 147
Zhou, Z., 4–9, 36
Zhu, C.Q., 144
Zhu, G.D., 120
Zhu, H., 145–146
Zhu, J., 136–137
Zhu, J.H., 136–137
Zhu, X., 46
Zhu, X.R., 131
Ziehr, B., 132
Zimmerman, L., 37
Zitzmann, N.U., 84
Zollo, M., 144–145
Zouboulis, C.C., 145–146
Zucchini, S., 135

SUBJECT INDEX

Note: Page numbers followed by "*f*" indicate figures, and "*t*" indicate tables.

A

Active pharmaceutical ingredients (APIs), 106
Activity cliff generator, 15
Activity landscape modeling, 14–15
Acute lymphoblastic leukemia (ALL), 128
Adsorbed collagen IV, 89
Alanine mutagenesis data, 44
Allene oxide synthases (AOLs), 114
α-Synuclein, 153–155
Alpha-domain proteins, 60–61
Alternative splicing factor (ASF), 136–137, 151–152
Alzheimer's disease (AD), 149–151
Ambiguous Interaction Restraints (AIRs), 37
Amyloid beta (Abeta) peptide, 29
Anaerobic respiration, 114
Antimicrobial peptides, 29
Apolar tunnel, 67
Apoptosis, 141, 143*f*
Apoptotic signaling cascade, 141
Arachidonic acid, 113–114
Association process, 61–62
Atomic force microscopy (AFM), 87
AutoDock, 18–19
AutoDock Vina, 15
Autophosphorylation, 128–129
5-Azacytidine, 2–3

B

Basement membrane (BM), 82–83
Binding affinity, 29–30
Binding free energy, 35–36
Binding motif, 28–29
Biotin synthesis, 112
Boltzmann's constant, 30–32

C

CABS dock protocol, 38
Carbon monoxide, 65–66, 65*f*
Cardiac troponin T (cTnT), 137–138
Catalytic cycle
- catalytic oxidation, 115
- difference spectra, 115–117
- electron donation, 115–117
- electrons transport, 115–117
- graphical representation of, 115, 116*f*
- iron–oxygen bond, 115–117
- iron transition, 115–117
- P450 mechanism, 115
- porphyrin ring, 115–117
- spin state, 115–117

Catalytic cysteine (Cys1226), 13–14
Cell–biomaterial interface
- adsorbed collagen IV, 89
- atomic force microscopy (AFM), 87
- cells, 87
- collagen IV
  - adsorption and assembly, 94–96, 95*f*
  - degradation, 90–91
  - HUVEC interaction with, 96–98, 97–98*f*
  - mixed NH2/CH3 SAMs, remodeling on, 98, 99*f*
  - quantitative studies, 94, 98–100
  - remodeling by endothelial cells, 91
  - reorganization and enzymatic remodeling, 92–93, 93*f*
- immunofluorescence
  - focal adhesion contacts, 88
  - integrins, 88
  - vinculin and FAK costaining, 88–89
- initial cellular interaction
  - cell adhesion and spreading, 88
  - cell morphology, 87–88
- model surfaces
  - mixed NH_2/CH_3 thiol SAMs, 86
  - self-assembled monolayers, 85–86
- morphological observations, 91–92, 92*f*
- protein adsorption, 86–87
- western blot analysis
  - FAK assay, 89–90
  - MMP-2 assay, 90

Cell–materials interaction, 83–84
Cerebratulus lacteus (CerHb)
apolar tunnel, 67
function of, 66
Gln44(E7), 67
Leu86, 67, 68*f*
neuronal mini-Hb of, 66
O_2 dissociation constant, 66
TyrB10–GlnE7 pair, 66
Cerebrospinal fluid (CSF), 154
ChEMBL, 4–9
CHEMBL3126646, 15, 21–22
Chemical information
assay ID number (AID), 4–9
epigenetic relevant chemical space (ERCS), 9
human epigenetic enzyme and modulator database (HEMD), 4
molecular databases, 4–9
representative confirmatory bioassays, 4–9, 7–8*t*
structure–activity relationships (SAR), 9
Chemoinformatic-based pharmacophore model, 12–14
Chemoinformatic studies. *See also* DNA methyltransferases (DNMTs)
activity landscape modeling, 14–15
chemoinformatic-based pharmacophore model, 12–14
epigenetic relevant chemical space (ERCS), 9–12
quantitative structure–activity relationships, 15
Chimeric proteins, 75–76
CH3 SAM, 96
CH_3 substrata, 94
Cineole, 110, 112*f*
Class I systems, 110
Class II systems, 110–112, 111*f*
Class III systems, 112
Class IV systems, 112–113
Class V systems, 113
Class VI systems, 113
Class VII systems, 113
Class VIII systems, 113–114
Class IX systems, 114
Class X systems, 114
Collagen IV
adsorption and assembly, 94–96, 95*f*
degradation, 90–91
HUVEC interaction with, 96–98, 97–98*f*
mixed NH2/CH3 SAMs, remodeling on, 98, 99*f*
quantitative studies, 94, 98–100
remodeling by endothelial cells, 91
reorganization and enzymatic remodeling, 92–93, 93*f*
Computational approaches, 4
Computer-assisted drug repurposing, 19–20
Confusion matrix, 13–14
Cspase-9 (Cas-9), 141–142
CYP108B1, 120–121
Cytochrome P450 Engineering Database (CYPED), 115
Cytochrome P450 reductase (CPR), 110–112
Cytochromes P450
background and history of, 106–109
catalytic cycle, 115–117, 116*f*
classes, 110–114
family nomenclature, 114–115
in industry, 119–123
uncoupling, 117–119

D

DC_05, 16, 17*f*
Decitabine, 2–3
Densitometry analysis, 89–90
l-Dependent potential energy function, 33–34
Diclofenac hydroxylation (CYP2C9), 122
Dideoxy DNA sequencing, 109
Dissociation process, 61–62
Distal effects, 61–62
Divinyl ether synthases (DESs), 114
DNA hypermethylation, 2
DNA methylation, 2
DNA methyltransferases (DNMTs)
chemical information, progress on, 4–9
chemoinformatic studies of
activity landscape modeling, 14–15
chemoinformatic-based pharmacophore model, 12–14
epigenetic relevant chemical space (ERCS), 9–12

quantitative structure–activity relationships, 15
computer-assisted drug repurposing, 19–20
epi-probes, 2
food chemicals as potential modulators, 20–21
representative compounds, 2–3, 3*f*
virtual screening (VS) hits
follow-up, 18–19
novel, 16–18
Docking, 16, 18
applications of, 21–22
Dopaminergic neurones, 155
Down syndrome (DS), 128
Down syndrome critical region (DSCR), 128
Drosophila melanogaster, minibrain in
abnormal neuroblast organization, 131
beta-tubulin, 131–132
cyclin-dependent protein kinases, 131
Dyrk1 substrates, 131
microtubule polymerization, 131–132
mutation, 131
nuclear factor of activated T cells (NFAT), 131
optomotor response, 131
phosphorylation, 131
RNAi ablation, 131–132
synaptic recycling, 131
Synj synaptic activity, 131
Drug repurposing, 19
Dynamical motions, 62
DYRK1A
in cancer
apoptosis, 141–145
cell proliferation, promoter of, 145–147
myeloproliferative disorders, 147–148
p^{53} family, 139–141
cardiovascular function of
aberrant calcium signaling, 136–137
alternative splicing factor (ASF), 136–137
calcineurin–FAT transcription factor, 137
calcineurin-responsive gene expression, 138
calcium/calmodulin-dependent protein kinase II isoforms, 136–137
cardiac hypertrophy, 138
cardiac troponins, 137–138
cardiomyocytes, 137
microRNAs, 138
myocardial infarctions, 136–137
serine/arginine-rich (SR) proteins, 136–137
Valsartan inhibition, 136–137
cell proliferation, promoter of
HPV16-infected cells, 146–147
Kaiso activity, 145
neoplasia, 145–146
Notch and Hedgehog signaling pathways, 145–146
REST, 146
RNA II polymerase kinase, 147
STAT3 inhibitors, 145–146
Wnt signaling pathway, 145
in infection and immunity
highly active antiretroviral treatment (HAART), 133
HIV-1 long-term repeat (LTR) activity, 132–133
HIV-1 Nef virulence, 133
HIV-1 replication, 134
inhibitor harmine, 134–135
Itk kinase activity, 134
lactones and chalcones, 135
macrophages, 132
neurodegenerative tauopathies, 135
NFkB, 135
protein kinases, 132–133
SNP genotyping, 133–134
Tau aberrant splicing, 135
T_{reg}/Th17 differentiation pathways, 134–135
unbiased chemical biology approach, 134–135
myeloproliferative disorders
acute megakaryoblastic leukemia (AMLK), 147
angiogenesis suppression, 147
globin transcription factor 1 (GATA1), 147–148
NFAT signaling, 147–148

DYRK1A (*Continued*)
Ts1Rhr trisomic murine model, 147–148
neuronal function
in PD, 155–157
synuclein regulator, 153–155
in Tau-related neurodegenerative diseases, 149–151, 150*f*
Tau splicing regulation, 151–153

E

End-point methods, 35–36
Epigenetic-related disorders, 21–22
Epigenetic relevant chemical space (ERCS)
bromodomains (BRDs) inhibitors, 10–12
data set, 9–10
Data Warrior, 10
DNMT1 inhibitors, 10, 11*f*
everything added to food in the United States (EAFUS), 10, 11*f*
generally recognized as safe (GRAS), 10, 11*f*
histone deacetylases (HDACs), 10–12
molecular diversity analysis, 9–10
molecular equivalence index (MEQI), 9–10
multiepigenetic therapies, 10–12
visual representation, 9–10
Epi-informatics, 4
Erythromycin N-demethylation (CYP3A4), 122
Eukaryotic mRNAs, 42–43
Exogenous ligands, 61–62
Exo-hydroxylation, 107
Exploratory analysis, 14–15
Extracellular matrix (ECM), 82–83

F

FAD-dependent ferredoxin, 112
FAK assay, 89–90
Ferredoxin, 110, 112–113, 122–123
FITC-Col IV, 94, 96
adsorption, 86–87
FlexPepDock, 37
Fluorescent assay, 90
FMN-containing flavodoxin, 113–114
FN fibrils, 92, 98
Focal adhesion contacts, 88
Focal adhesion kinase (FAK), 83–84
Food chemicals, 20–21
nearest-neighbor curves, 20–21, 21*f*
systematic mining of, 20–21
Forkhead in rhabdomyosarcoma (FKHR), 142
Free-energy calculations, 74–75

G

GalaxyPepDock, 37–38
GalaxyTBM, 37–38
Genomic data, 122–123
Glide, 18
Gln44(E7), 67
Global docking, 38–39
Globins, structural plasticity in
heme group, coordination of, 73–76
ligand migration pathways
Cerebratulus lacteus mini-Hb, 66–67
hemoglobin, 65–66
myoglobin, 63–65
protoglobin, 72–73
truncated Hbs, 67–72
structure–dynamics–function relationships, 62–63
G-protein-coupled receptors (GPCRs), 43

H

HADDOCK protocol, 41
HEK293 human embryonic kidney cell, 156–157
Heme group, coordination of, 73–76
Hemoglobin
α and β chains, 65–66
carbon monoxide, 65–66, 65*f*
ligand migration in, 65–66
R- and T-states, 65–66
High Ambiguity Driven protein-protein Docking (HADDOCK), 37
High-throughput screening, 123
ligand databases, 45
HisE7 gate hypothesis, 63–64, 64*f*
Histidine F8 (HisF8), 60–61
Histone modification, 41
Human DNMT1 (hDNMT1), 12, 18
Human down syndrome kinase DYRK1A
in cancer, 139–148
cardiovascular function of, 136–138

drosophila, minibrain in, 129–132
in infection and immunity, 132–135
multifunctional role of, 128–129, 129*f*
neuronal function of, 148–157
Human embryonic kidney (HEK293), 144
Human parathyroid hormone, 44–45
Hydroperoxide lyases (HPLs), 114
Hydrophobic CH_3 surfaces, 85–86
Hydrophobic pocket, 64
Hydroxylations, 106
Hypomethylating agents, 19–20

I
ImageJ plug-ins, 88
Immunofluorescence
focal adhesion contacts, 88
integrins, 88
vinculin and FAK costaining, 88–89
Initial cellular interaction
cell adhesion and spreading, 88
cell morphology, 87–88
in silico techniques, 4
Integrins, 88
Interatomic interactions, 28
Iron–sulfur cluster, 113

J
JNK-ASK1 proapoptotic pathway, 144

L
Leu86, 67, 68*f*
Lewy bodies (LB), 153–154
Ligand-docking sites, 74
Ligand migration pathways
Cerebratulus lacteus mini-Hb, 66–67
hemoglobin, 65–66
myoglobin, 63–65
protoglobin, 72–73
truncated Hbs, 67–72
Local docking, 36–38
Locomotor defects, 155
Lysine-specific demethylase (LSD1), 42

M
Major histocompatibility complex (MHC), 36
Matrix metalloproteinases (MMPs), 83
MedusaDock, 38
Mesencephalic DA (mDA) neurones, 155–156
Metastatic progression, 144
Methylenedisalicylic acid, 18–19
Mevastatin, 119, 119*f*
miR-1246, 140
Mixed NH_2/CH_3 thiol SAMs, 86
Mixed SAMs, 95–96, 95*f*, 97*f*
MMP-2 assay, 90
Modeling key interactions, 43–45
Molecular mechanics/generalized Born surface area (MM-GBSA), 35–36
Molecular mechanics/Poisson–Boltzmann surface area (MM-PBSA), 35–36
Molecular operating environment (MOE), 13–14
Molecular oxygen, 106
Molecular switch, 70
Monoclonal anticollagen IV antibody, 89
Myoglobin
HisE7 gate hypothesis, 63–64, 64*f*
ligand migration, 63
oxygen migration, 63–64
Trp mutation, 64

N
Native Col IV remodeling, 98–100, 99*f*
Neurofibrillary tangles (NFTs), 149–151
Neuroglobin (Ngb)
chimeric proteins, 75–76
CysCD7, 74
CysD5, 74
free-energy calculations, 74–75
hexacoordination equilibrium, 75–76, 75*f*
highly oxygen-consuming tissues, 74
ligand-docking sites, 74
molecular simulations, 74–75
6c and 5c globins, 74
3-over-3 globin fold, 74
tunnels, 72*f*, 75–76
NH_2 substrata, 92
Nitric oxide dioxygenase (NOD) activity, 62, 70
Nonequilibrium approaches, 33
Nuclear factor I/B (NFIB), 140

O

O_2 dissociation constant, 66
Omeprazole hydroxylation (CYP2C19), 122

P

Packing defects, 62–63
Parkinson's disease (PD), 148–149, 155–157
Partial trisomy, 148–149
Pathway free energy methods
  in alchemical approaches, 33–34
  decoupling transformations, 34
  FEP approaches, 34–35
  partition functions, 32–33
  in physical approaches, 33
  second transformation, 33–34
  thermodynamic cycle, 33–34
  unbound *vs.* bound states, ligand-protein complex, 32–33
P450cam, 107, 108–109*f*, 109
PD. *See* Parkinson's disease (PD)
PepATTRACT, 39
Peptidomimetic drugs, 43–44
Pericellular proteolysis, 83
Pharmacophore, 12
Phe(145)G8, 71*f*, 73
Porphyrin ring, 106–108
Pravastatin, 119–120, 119*f*
Predicting protein–peptide interactions
  conformational flexibility, 39–40
  discrete molecular dynamics (DMD), 39–40
  HADDOCK protocol, 40–41
  hierarchical clustering, 39–40
  molecular dynamics (MD), 39–40
  root mean square deviation (RMSD), 39
  troponin C *vs.* end terminal helix of troponin I, 39
P450RhF, 122
Prostacyclin synthases (PGISs), 114
Protein adsorption, 86–87
Protein Data Bank, 36
Protein dimerization, 73
Protein–ligand interaction fingerprints (PLIF), 13–14
Protein–peptide binding affinities
  BEDAM method, 48–49
  canonical binding site, 49
  cytoplasmic FMRP interacting protein 1 (CYFIP1), 48–49
  double-decoupling alchemical approach, 48
  fragile X mental retardation protein (FMRP), 48–49
  free energies, 47–48
  HADDOCK program, 49
  HIV integrase, 49–51, 50*f*
  PMF approach models, 48
  protein–drug binding, 49–51
  SH2 regulatory domain, 47
  Src homology 2 (SH2), 47–48
Protein–peptide docking
  global docking, 38–39
  local docking, 36–38
Protein–peptide interactions (PPIs)
  sample applications
    binders from nonbinders, 45–47
    binding affinities, 47–51
    modeling key interactions, 43–45
    molecular dynamics, 41–43
    predicting interactions, docking, 39–41
  theory and methods
    MM-PBSA, MM-GBSA, 35–36
    pathway free energy methods, 32–35
    software, 36–39
    statistical mechanics formulation, 30–32
Protoglobin (Pgb)
  dual-path ligandmediated migration mechanism, 73
  heme moiety, 72
  ligand-dependent equilibrium, 73
  ligands, 72–73
  O_2 dissociation rate constant, 72–73
  Phe(145)G8, 71*f*, 73
  protein dimerization, 73
p^{53} tumor suppressor protein, 139–140
PubChem, 4–9, 7–8*t*
Putidaredoxin, 110, 120–121

Q

Quantitative structure–activity relationships, DNMTs, 15

R

Raman spectroscopy, 107
Rb/E2f signaling, 137
Receiving operating curve (ROC), 13–14
Redox partner, 121
Redox proteins, 122–123
Replica Exchange Monte Carlo methods, 38
Representative computational studies, 4, 5–6*t*
RNA recognition motif (RRM), 152

S

S-adenosyl-L-homocysteine (SAH), 15–16
S-adenosyl methionine (SAM), 2
SAH. *See* S-adenosyl-L-homocysteine (SAH)
Salvador–Warts–Hippo pathway, 142–144
SC35, 153
SEPT4, 156–157
Serine/arginine rich (SRp55), 153
Similarity principle, 14. *See also* Activity landscape modeling
Single-decoupling alchemical approaches, 34
siRNA kinome, 149–151
Sirt1-mediated deacetylation, 142
SNAIL-derived peptides, 42
Soluble matrix proteins, 84–85
S129 phosphorylation, 154–155
SR proteins, 152
Standard numbering schemes, 115
Stapled peptides, 42
Statistical mechanics formulation
- binding affinity, 30–32
- configurational partition functions, 30–32
- free energy perturbation (FEP) approaches, 32

Streptomyces carbophilus, 110–112
Structure–activity information (SAI), 4
Structure–activity similarity (SAS) maps, 14
Structure–dynamics–function relationships, globins, 62–63
Surflex-Dock, 15

T

Tanimoto coefficient, 14–15, 19–20
Tau exon 10 (E10), 151–153
Tau isoforms, 151–152
Tau-mediated NFTs, 149–151, 150*f*
Tau-related neurodegenerative diseases, 149–151
Tau splicing regulation, DYRK1A, 151–153
Thermodynamic integration (TI), 43–44
Threonine 522 (T522), 142
Thromboxane synthases (TBAXs), 114
Trp mutation, 64
Truncated Hbs
- dual-path ligand-induced dynamical mechanism, 68–69, 69*f*
- long tunnel (LT), 68–71, 71*f*
- nitric oxide, 68–69
- NOD activity, 67–68
- protein backbone, dynamical motion of, 70
- residue G8, 71–72
- STE7, 67–68, 70–71
- STG8 tunnel, 68–69
- 2-over-2 helical motif, 67–68

Type-b heme, 106, 107*f*
Tyr^{96}, 108–109
TyrB10–GlnE7, 66
TyrB10–GlnE11, 70

U

Uncoupling. *See also* Cytochromes P450
- catalytic turnover, 118
- coupling efficiency, 117–118
- cysteine pocket, 117–118
- molecular oxygen, 117
- phenylalanine residue, 118
- push effect, 117–118
- shunt pathways, 117–118
- substrate binding, 117–118

USA Food and Drug Administration (FDA), 2–3

V

Vinculin and FAK costaining, 88–89
Virtual screening (VS) hits
- follow-up, 18–19
- novel, 16–18

Vitamin D_3
- bioconversion of, 120
- two-step hydroxylation, 120, 120*f*

W

Water contact angle (WCA), 86
Western blot analysis
  FAK assay, 89–90
  MMP-2 assay, 90
Whole-cell patch-clamp assays, 44

X

X-ray crystallographic data, 44–45

Z

Zymograms, 94
Zymography, 90–91